PAJ Books
Bonnie Marranca and Gautam Dasgupta
Series Editors

Art + Performance
Meredith Monk, edited by Deborah Jowitt
Rachel Rosenthal, edited by Moira Roth
Reza Abdoh, edited by Daniel Mufson
Richard Foreman, edited by Gerald Rabkin

Art + Performance

Richard Foreman

Edited by Gerald Rabkin

A PAJ Book

The Johns Hopkins University Press   Baltimore + London

© 1999 The Johns Hopkins University Press
All rights reserved. Published 1999
Printed in the United States of America on acid-free paper

9 8 7 6 5 4 3 2 1

The Johns Hopkins University Press
2715 North Charles Street
Baltimore, Maryland 21218-4363
www.press.jhu.edu

Library of Congress Cataloging-in-Publication Data

Richard Foreman / edited by Gerald Rabkin.
    p.   cm.—(PAJ books. Art + performance)
  Includes bibliographical references.
  ISBN 0-8018-6113-6 (alk. paper).—ISBN 0-8018-6114-4
(pbk. : alk. paper)
    1. Foreman, Richard, 1937–   —Stage history.   2. Theatrical
producers and directors—United States—Interviews.
3. Dramatists, American—20th century—Interviews.
4. Foreman, Richard, 1937–   —Interviews.   5. Experimental
theater—United States.   6. Drama—Psychological aspects.
7. Consciousness in literature.
I. Rabkin, Gerald. II. Series.
PS3556.O7225Z87 1999
812'.54—dc21   98-49863   CIP

A catalog record for this book is available from the British Library.

Permissions may be found on pages 247–48.

Frontispiece: Richard Foreman on the rehearsal set of *Paradise Hotel* (1998). Photograph by Paula Court.

To Masha

Contents

Photograph gallery follows page 142.

Acknowledgments

I am grateful to my long-time associates Bonnie Marranca and Gautam Dasgupta for their invitation to be part of this series. I'd also like to thank the staff of the Johns Hopkins University Press, especially Tom Roche, whose copyediting has been assiduous and perceptive, and Linda D. Tripp (no, not *that* one), the former Humanities Editor, who proved most helpful and dependable. My wife, Diane Jacobs, offered, as always, a detailed critical reading of my own contribution. Many thanks as well to the photographers Babette Mangolte and Paula Court, the pictorial chroniclers of Foreman's early and late work, respectively. And much gratitude to Richard Foreman himself, who not only provided me with invaluable material, but who was never too busy for just one more question.

Richard Foreman

Gerald Rabkin

Richard Foreman: An Introduction

So silence and babble, so both, so linked, so opposed and coordinated, in that new, next attached level of babble, laid forth blanket-like under the light, always the light re-showing itself outside which was only a way, finally, of being inside.

Samuel II

Why strings and decorated glass walls between the stage and the audience? Why disruptive noise and light? Because somebody (who?) lusts, above all else, for the evocation of that dense babble of signs and energies out of which normal, everyday life, surfaces as the transitory, heartbreaking thing it really is.

Program, *Permanent Brain Damage*

In a 1993 interview about the future of avant-garde theater, Richard Foreman looked back wistfully: "It's difficult to imagine having a career [now] doing the kind of thing people like me have done throughout the years. . . . When I began at the end of the '60s, there was the genuine illusion of a genuine counterculture. That illusion has dried up."[1] American experimental theater is fortunate in the timing of Richard Foreman's career. Born in New York City in 1937 but raised in Scarsdale, he returned to the city in 1962, with a B.A. degree from Brown and an M.F.A. degree from Yale, to pursue a conventional playwriting career. But he reentered the city as a new cross-disciplinary avant-garde was ascending, partially in response to political imperatives. The early 1960s saw the heyday of Happenings and the performative art of the Fluxus group, the maturation of the Living Theatre, the increasing importance of new café theaters (Cino, La MaMa), leading to a new Off-Off-Broadway alternative; the period saw Cage, Cunningham, Warhol

1

and other pop artists, underground film, the beginnings of postmodern dance at the Judson Church—all signs of a new, visionary energy that was to swell as the decade became increasingly politicized and the climate of dissent intensified.

Foregoing genuine commercial prospects, the aspiring young playwright felt himself consumed by this experimental energy. In the latter years of the decade, he was ready to add his own contribution. In 1968, the year Richard Foreman founded the Ontological-Hysteric Theater, the "genuine counterculture" that he alludes to above was at its apogee, particularly in the performing arts. The year saw the Radical Theatre Festival's gathering at San Francisco State College of street and guerrilla theaters; the Ridiculous theatres' productions of *The Moke-Eater* and *Turds in Hell;* the premiere of the Open Theater's *The Serpent* in Rome; and the debut of Richard Schechner's Performance Group with *Dionysus in '69.* In London, Peter Brook directed an innovative *Tempest,* a piece that contained the seeds of his subsequent radical work in Paris; Grotowski's Theatre Laboratory triumphed from Edinburgh to Aix-en-Provence and was soon to triumph in the United States. The modern black theater movement effectively started in 1968, with the founding of the Negro Ensemble Company. And recall the impact of Joseph Papp's Public Theater, which followed its debut offerings of *Hair* and *Hamlet* in 1967 with Václav Havel's *Memorandum* in 1968. And, 1968 saw the triumphant, if controversial, return from European exile of the Living Theatre, which filled auditoriums, gymnasiums, churches, and theaters from Berkeley to Brooklyn.

But if 1968 was an *annus mirabilis* for radical performance, it was also, in a way, a last hurrah. The triumphs of a community of dissent were soured by less auspicious events: the crushing of student rebellions in Mexico City, Chicago, and Paris; the assassinations of Dr. King and Robert Kennedy; the Soviet extinction of the Prague Spring. Activism abated, and countercultural unity dissipated into individual agendas, political and artistic. As the barricades fell, experimental theater increasingly turned away from political commitment toward more introspective concerns. Foreman's first production, *Angelface,* was a harbinger of this change. In style and theme it differed radically from other sixties experiment: the piece's opaque language, its accumulation of mysterious images without social referents, its inward turn toward questions of consciousness and perception, the very name of its theater—Ontological-Hysteric—suggested that something new was afoot.

Soon, Foreman would be joined (albeit in their own individualistic ways) by others with similar affinities, preeminently, Robert Wilson and the members of the Mabou Mines collective, a disparate coupling that Bonnie Marranca was to subsume under the rubric "Theatre of Images." In 1968 Wilson was conducting dance / theater experiments with the Byrd Hoffman School of Byrds in preparation for his first major work, *The Life and Times of Sigmund Freud*, performed the following year; and Mabou Mines was founded the year after that, in 1970.

The emergence of Foreman and Wilson, some have argued, began a period in which American experimental theater became depoliticized. This tendency to read Foreman's obsessional subjectivity as a rejection of *all* politics, is, in my view, an error. Foreman's formal strategies, his solipsism and phenomenology, may obscure but do not reject a consistent rebellious stance. He has stated unambiguously that the principles of his theater have always put it ideologically "in direct opposition to the mythos of mainstream, business-oriented culture." Serious art "functions as an adversary to the going culture." That is why the powers-that-be consider it subversive. "It *is* subversive. It implies that the cultural choices we have made are wrong."[2] In this conviction he remains faithful to the values of the generation that spawned him.

Building a Foundation

Foreman's roots in the downtown arts community directed his artistic decisions. In starting his theater, he did not search out a conventional playhouse but borrowed the ground floor loft of Jonas Mekas's Cinemateque on Wooster Street in the neighborhood just then becoming known as SoHo. Nor did he seek out conventional actors; his performers were mostly avant-garde writers, filmmakers, painters, or their friends. In this supportive context, Foreman's theater made its debut with *Angelface*. Although this strange "play" contained named characters—Karl, Walter, Max, Rhoda—in seemingly ordinary domestic situations, its dialogue, stripped bare of normal social interaction, made no conventional sense. Foreman acknowledges that even the cognoscenti in the audience were bewildered. "For the first six years we knew that half our audience would walk out before the end of the play, often within the first twenty minutes."[3] When I saw Foreman's *Evidence* in 1972 at the old Theater for the New City on Bank Street, there was a sign that read, "If you leave during the production, please leave quietly." By the end of the piece, there were three of us left.

What was so disturbing? It was not merely the absence of conventional narrative and characters, for experimental audiences were used to this in the work of the Open Theater and other groups. But the innovative work of Chaikin, Schechner, and Grotowski affirmed strategies that Foreman actively rejected: thematic unity, ensemble creation, performative presence, and emotional commitment. In their place, Foreman substituted their opposites: thematic fragmentation, auteurish creation, writerly presence, and emotional estrangement. In this last regard, he carried the *Verfremden-Effekt* of Brecht, his "god" at the time, to a level of extremity. He arrived early at an aesthetic to which he has remained true: to purge art of emotional habit, through both the radical randomness of his drama and the jarring style of its staging (buzzers, noise, snatches of music, provocative lighting, etc.). Paradoxically, he found in theater—the ultimate public art—a way to attack the group mind. "When I make theatre, I prefer to think I'm directing it toward the individual, sitting in isolation from everyone else in the audience."[4] By exploring the only thing he really knew—his own mind—he meant to shock others into acts of recognition.

Another characteristic links Foreman with Brecht: they are the most intellectual of playwrights, enormously well read and discursively prolix. It is tempting to read Foreman's theater work primarily in terms of his acknowledged intellectual influences. And they are many: Gaston Bachelard provided the idea of a psychology that wasn't family history but was, rather, the individual's encounter with the physical givens of his environment; Gertrude Stein pioneered the sense of writing as a state of continual presence and the notion of always "beginning again"; the phenomenologists Husserl and Heidegger demonstrated the impenetrability of objects; the French structuralists and poststructuralists revealed the ubiquity of the signs and cultural codes within which we all function. One could go on.

But just as Brecht was theatrically more practical and empirical than his theory of epic theater would suggest, Foreman's theater is no mere articulation of theoretical ideas. Obviously, theory illuminates his plays, but to reduce his work to intellectual propositions would impose a transparency that Foreman rejects. His theater has sustained itself through the years because, at its best, it is *fun*—spirited, frenetic, dazzling, surprising, provocative, even as it is intellectually demanding. Foreman is not only an obsessive self-scrutinizer, a philosophical prober of recalcitrant reality, but also a ring-master melding text, performance, music, sound, and objects into unique spectacles.

The dual nature of Foreman's work is reflected in the name he gave his group: the Ontological-Hysteric Theater. The name derives, he explains, from "the basic syndrome controlling the structure" of his early plays, "that of classic, middle-class, boulevard theatre, which I took to be hysteric in its psychological topology."[5] Indeed, the early plays invariably begin with some sort of mysterious, repetitive hysterical conflict. And the ontological-hysteric dialectic suggests other polarities as well. Ontology, as beginning philosophy majors should know, is the branch of metaphysics that deals with the ultimate nature of being. The ontological nature of Foreman's theater resides in the fervent if unachievable quest for ultimate meaning: hence the fierce concentration on the reality of each individual moment, which Foreman often stops, like Gertrude Stein, to replay and rescrutinize. *Hysteria,* a more common word, is a psychological condition characterized by emotional excitability, excessive anxiety, and sensory and motor disturbances. Its sexist derivation from the Greek word for *womb* notwithstanding, the word evokes physical vulnerability, terror, and sexuality, an irrational counterbalance to the implacable rationality of Foreman's mental processes. When Kate Manheim, Foreman's muse and, later, wife, joined the Ontological-Hysteric Theater for its fourth production, *Hotel China,* this dialectic intensified through her sensual but quizzical presence.

In a theatrical sense, the juxtaposition of ontology and hysteria reflects the relationship between Foreman the playwright and Foreman the director.[6] Philosophy is—or should be—a rational quest, and Foreman's plays deny empathy, in part because he does not want to pollute reason. On the other hand, theater is nothing if not corporeal, and in the state of hysteria the claims of the body swamp the mind. That is why Foreman chose the theater as his main medium and not fiction, poetry, or discursive prose. Theater is here, now, palpable.

In the earliest plays, Foreman the director consciously served Foreman the playwright. "The sets were extremely minimal. . . . I only built those elements of the set that the play made specific reference to. . . . I built everything. But I added no decoration. . . . I wanted . . . the look of my awkward carpentry."[7] But by the time of *Hotel China* (1971–72) Foreman's interest, enhanced by his readings in phenomenology, began to shift from his scrutiny of the psychological subtexts that underlie the surface of conscious action; he now became fascinated with how mental registerings collide with the impenetrable objects we encounter in the world. He began to imagine intricate, strange objects that would suggest how things in general manipulate us. And so Foreman the designer assumed a greater role in his self-collaboration. He increasingly felt that a minimally dressed space would not allow the text to ricochet fully between levels of meaning. He saw the stage not, as conventional directors do, as a platform on which to display action, but as "a reverberation chamber which amplifies and projects the music of the action so it can reveal the full range of its overtones."[8] All the materials of theater—scenery, props, lights, noises, bodies—should be thrown together in polymorphous play.

Take, for example, the opening of *Hotel China:* There are two naked bulbs on floor stands, and two men sit behind each bulb. One has a sack over his head. A lamp is placed on top of his head over the sack and a wrapped package is placed upon the other man's lap. Blackout. When the lights come up again a rock has replaced the package. The unhooded man says, "Get this rock off." The hooded one replies, "I didn't notice it." As the scene continues, there is another blackout and the sound of a rock being hit by a hammer. When the lights come back up, the men are gone and a rock, gently swinging, is suspended in the air illuminated by a flashlight. The men return. Signs are placed on the rear wall, one of which reads: EACH ROCK IS CAPABLE OF VIBRATING INDEPENDENTLY. A screen descends; a film is shown, consisting of shots of rocks being placed in different positions in different rooms. The film goes to white, the projector still running, as a big six-foot rock is rolled on stage and keeps rocking. Bird chirps are heard and continue until the rock comes to rest. A single bright floodlight illuminates the stage. The bird songs fade as the crew places small rocks around the big rock. There is the noise of a hammer striking rock repeated ten times. Silence.

There is indeed an echo of Beckett here, but it is consciously distorted. The rocks are "characters" in the piece, as significant as the human actors. One visual element which makes its first appearance in *Hotel China* came, by subsequent repetition, to be synonymous with Foreman's theatrical style: In one scene, he attached strings to the sides of the stage in a widening funnel, which came to a point on an actor's brow. The strings suggested lines of force emanating from the character's—the author's—yearning mind. In the later plays, these ubiquitous strings, intensified by repetitive black dots, developed a life of their own. Feeling that theatrical space is insufficiently defined, Foreman wanted the strings, together with lights focused into the audience's eyes, to turn the stage into a participant-provocateur in the play, rather than have it remain a neutral site of performance.

Midstream

In *Rhoda in Potatoland* (1975–76), the eponymous heroine (played by Kate Manheim) is asked by a crew person (played by Foreman): "Aren't you the famous Richard Foreman?" She smilingly answers, "Yes," and he continues: "It's an honor to meet you. . . . See? You're famous."[9] Here, Foreman ironically notes not only the increasing attention that he and his theater began to receive in the early seventies but also the ascendancy of Kate Manheim as his artistic and personal partner. What might be termed Foreman's middle period (roughly from 1974 to the early eighties) is defined by Kate's central role in his life and work, her now dominant presence in his theater as both artistic surrogate and object of desire.

But before considering this crucial collaboration we must examine a vital decision that also defines this period: the Ontological-Hysteric Theater assumed residency in a space Foreman had bought at 491 Broadway. This move to a permanent (for a while) performance space inevitably influenced Foreman's scenic choices, because the space was an eccentric loft, only 14 feet wide. Converted into a theater, it provided room for only seven rows of seats, which occupied less than a quarter of the space. The stage itself was a full 75 feet deep, the first 20 feet of which were at floor level. The next 30 feet ran at a steep rake, with the stage finally leveling off at about a 6-foot height for the remaining depth. Foreman used this eccentric space to enhance his productions: during performances actors and objects would often roll down that 30-foot rake, and sliding walls would enter from the side of the stage, creating a series of quickly changing areas that varied radically in depth. The stage floor was built from scratch and changed for each production; it could slope up and down like a series of hills, or a wall could move in and produce a shallow space. Throughout each show the performance areas would alternately close and open up, shifting back and forth vertiginously. To this observer's eye it was the most hospitable space Foreman ever worked in, the perfect site for his theater of disequilibrium.

Four major plays were produced in this space: *Pandering to the Masses: A Misrepresentation* (1975), *Rhoda in Potatoland* (1975–76), *Book of Splendors: Part II* (1977), and *Blvd de Paris: I've Got the Shakes* (1977). A play called *Madness and Tranquility* was presented in open rehearsals in 1979, but the production was canceled. At this point, Foreman stopped to assess his artistic options and his future. He could continue comfortably at 491 Broadway with his faithful audience, but this seemed too predictable, too safe. So he decided to sell the building and move on. In an essay titled "The Decline and Fall of the (American) Avant-Garde," Richard Schechner dates the end of the heyday of American experimental theater (which he begins with Cage and Cunningham in 1952) with the closing of Foreman's theater on lower Broadway in 1979.[10]

In all of the above plays, Kate Manheim played the leading role of Rhoda.[11] Foreman maintains that he never wrote material specifically for Kate, that he wrote as he always did with his usual preoccupations. But he acknowledges being aware that Kate would now be the pivotal figure in his work: "Kate's planet had entered my universe, and the gravitational force of that planet certainly pulled me in new directions."[12] Certainly in new erotic directions. Her body—often presented nude—became a physical presence that the philosophic mind could not ignore. Seeing was not only the means of knowledge but also the pathway of desire.

Rhoda had been one of Foreman's stock company of surrogates from the very beginning. Kate assumed the role in *Sophia = (Wisdom): Part 3: The Cliffs* (1972–73), and from this point her centrality steadily increased. In the earliest plays, no character has more weight than any other, nor indeed than the objects that entrap them. But Rhoda / Kate now separates herself from all the rest. If she possesses a weighty antagonist, it is Foreman himself, present through the sepulchral tones of his taped voice or his invisible hands on the audio controls.

In the Rhoda plays, Rhoda / Kate is often harassed and tormented (e.g., from *Rhoda in Potatoland*: "Rhoda is left alone with a hooded man who appears in the chair. A table is set. She climbs on it, crying out in fear . . . ready for unimaginable tortures.")[13] But she is also volubly curious and cheeky, querying everyone and everything in her path, giving more than she gets. As Marc Robinson notes, "Rhoda is Foreman's savviest quester, willing and more than able to lead him out of suffocating quotidian life."[14] So the images of female bondage and passivity (at one point in *Potatoland*, Rhoda holds up a book of Victorian erotica) give way to a more independent portrait whose nakedness is now a matter of personal choice.

Rhoda's nakedness suggests to Foreman a new metaphor, one that he explores obsessively. In *Rhoda in Potatoland* his taped voice intones: "The naked body as a vast space. Travel in it. To travel in it is to be in a landscape that you KNOW how to relate to."[15] Which implies the reverse: to travel outside it is to be in a landscape you don't know how to relate to. Travel—by assaulting us with language and customs that bewilder us—forces us to recognize what Foreman has known all along: consciousness demands confronting the strange. We are all tourists in the terra incognita of reality. But no matter how we are imperiled, only by becoming a tourist can we escape the inertia of immobility. Rhoda, of course, is the tourist par excellence, engaged in a continuing quest that reflects her restless need for self-expression, a quest both exterior and interior.

The fantasy of Potatoland was soon to yield to more recognizable foreign turf. Foreman's growing international reputation had brought him an invitation to the prestigious Festival d'Automne in Paris, where, in 1976, he presented *Livre des Splendeurs* at Les Bouffes du Nord. Thus began a warm artistic relationship between Foreman and the French that has lasted through the decades. In a 1983 interview he remarked: "When I first came here [Paris] in 1959 . . . I fell in love with the city. New York I find so physically oppressive. Spiritually, I could never stop working in New York, because my dialogue is with my own culture. But in Paris the theater community is much more open to ideas."[16] On returning to New York, his dual allegiance found expression in *Blvd de Paris: I've Got the Shakes* (1977). Here journey and disequilibrium were the basic motifs: fragments and images of travel—traffic clamor, conveyances, shifting lights and screens, suitcases, and so on—reinforced the central theme: "She goes back and forth between parts of herself."

This equation of travel and disequilibrium was to outlive both the lower Broadway theater and the character of Rhoda. As late as 1983, in *Egyptology* (presented at the Public Theater), Kate is still the protagonist, though she plays not Rhoda but an unnamed aviator who crashes in a foreign land, now more third-worldish than French. At least two other figures in the play are beleaguered not-so-innocents abroad who, finding themselves engulfed by exoticism, nonetheless discover signs and codes from the culture that spawned them. Foreman intensifies his images of dislocation ("You go from here to there. It doesn't work out."),[17] to such an extent that *Egyptology* veers more toward the social than any of his previous work. Not, of course, overtly; but Foreman seems genuinely overwhelmed by a vision of approaching apocalypse, as themes of revolution, poverty, imperialism, nuclear destruction, almost overshadow his personal and phenomenological concerns.

Other Voices, Other Rooms

As Foreman's reputation spread beyond the downtown arts community, new theatrical opportunities presented themselves. Institutional theaters came calling with increasing frequency (particularly, as his status grew, in Europe). In 1976 he received an offer from Joseph Papp to stage a revival of Brecht and Weill's *Threepenny Opera* at the Vivian Beaumont Theater in Lincoln Center (then under New York Shakespeare Festival management). Foreman had never staged the work of any playwright other than himself, but his affinity with Brecht was deep; in his production he worked to recover the savagery that Marc Blitzstein's popular Off-Broadway success had sacrificed. Even mainstream critics like Clive Barnes (then of the *Times*) praised the "clear and fierce power" of his direction, while, curiously, the *Village Voice* found the production *too* tough and austere.[18]

Raul Julia gave one of his most effective performances as a cold, sinister Macheath. I particularly recall a vivid image that occurred during the staging of the mordant, ironic "Barbara-Lied," in which Polly sings of the men who have wooed her. To those clean, neat, respectful, and respectable men she said, "No." But then along came Macheath, who was none of these things, and she couldn't say no. As Polly sang, there was absolutely no movement by her or Macheath: she faced the audience from center stage, with the elegant Macheath, back to us, staring at her from the center of the apron. Not a gesture from him, not a movement—until the last verse, during which he oh-so-slowly removed, finger by finger, his white gloves; he impassively placed them at Polly's feet as she completed the song and her surrender.

Threepenny Opera was the first of eight Foreman productions under Papp's auspices over the next dozen years. In addition to *Threepenny Opera,* Foreman directed four more plays written by other playwrights, and he directed three plays written by himself in the Ontological-Hysteric tradition, *Penguin Touquet, Egyptology,* and *What Did He See?* (The relationship with the Public Theater was revived after Papp's passing with the 1996 production of Suzan-Lori Parks's *Venus*.)[19]

In approaching the work of others, Foreman's search for creative affinities often leads him to use his trademark visual vocabulary. In his production of *Don Juan* in Central Park in 1982, for example, the set—an elaborate decaying wall lit by huge chandeliers that rose and fell—was covered by the familiar intricate network of strings, which served here as the grid of interpretation that must be used to recover the past. The production was a revision of the staging of the play that Foreman had done for the Tyrone Guthrie Theatre in Minneapolis the previous year, an event important enough for Frank Rich to fly out to cover it for the *New York Times.* Rich pointed out that this version of Molière looked a lot like *Penguin Touquet,* which had played the previous year at the Public Theater: "The Guthrie auditorium is bisected by wires; fractured alphabets hang from some of them. The lighting is often harsh and blinding, the sound effects . . . thunderous. . . . there is no mistaking this *Don Juan* for the work of any other director."[20]

Two of the plays Foreman staged for the Public Theater were contemporary European plays: Botho Strauss's *Three Acts of Recognition (Trilogie des Wiedersehens)* in 1982 and Václav Havel's *Largo Desolato* in 1986. The former, set in an art gallery where an exhibition called CAP- ITALIST REALISM is on private display, ironically scrutinizes the failings of the cultural and economic elite in a cautionary parable of bourgeois decay. Havel's play—which, of course, antedates the fall of Communism in Czechoslovakia—has as its protagonist a professor who has published a book that has disturbed the authorities and made him liable for arrest. His dissenter friends want *him,* not themselves, to be a hero; the government wants to make him an unperson. He is trapped between deceitful or impossible choices. Indeed, his head is a sledgehammer and he's got the shakes. Foreman seizes this anxiety and uses his familiar vocabulary: glaring lights, insistent buzzers, repetitive stylized movement, to help depict a society polluted on all levels by fear.

The two most controversial productions of Foreman's career occurred when he turned his hand to opera. This was not as unusual a move as it might on the surface seem, for Foreman had participated in a number of music-theater collaborations with Stanley Silverman since *Elephant Steps* at the Berkshire Music Festival in 1968. (Silverman also served as musical director for *Threepenny Opera.*)

In the early eighties Foreman received an invitation from the Paris Opera to stage *Die Fledermaus,* a recognition of the high cultural regard in which he has been held in France. But if the French cherish the avant-garde in general, they are also not reluctant to be outraged by it. The production opened in March 1983 to a firestorm of abuse. The account in *Variety* speaks for itself:

> Richard Foreman, the "underground" New York director, has been causing acute displeasure on both sides of the footlights with his freakish staging of *Die Fledermaus* at the Paris Opera. First-nighters shouted to turn off the reflectors that blazed on the house and blinded their eyes during a bacchanal interlude in which dancers mingled with black rollerskaters. The pro-

longed booing drowned out the orchestra. . . . Reviews have echoed the public disapproval of the production with unaccustomed ferocity. "Ugliness is everywhere—in the dryness of the decor, the vulgarity of the colors, and in the brutal lighting," complained Jean Cotte in *France Soir.* The critic likened Foreman to a wingless bat—a rat from New York's pseudo-intellectual cellars—incapable of making *La Chauve-Souris* take flight. . . . Due to the directorial black magic, old Vienna seemed to have moved to off-Broadway on amateur night.[21]

In our jaded world it is indeed unusual to infuriate the bourgeoisie in time-honored avant-garde style, but Foreman (with a little help from his friends) did it again in the unexpected venue of the Brooklyn Academy of Music's Next Wave Festival. *Birth of the Poet* was subtitled "an opera," but, though it indeed had a score by Peter Gordon, it was an opera only in the widest, experimental use of the word. First conceived as a collaboration between Foreman and punk novelist Kathy Acker, it enlisted the help of Gordon and painter David Salle. After tryouts in Rotterdam in 1984, the work was presented, with revisions, in December 1985 in Brooklyn. Acker's text, replete with obscenities, presented a lurid, eroticized vision of three barely related periods of world history: the late Roman Empire, contemporary Iran, and New York in a near-future apocalypse. Most agreed that the basic problem with the production was that its Cage-Cunningham aesthetic model—in which participating artists essentially work independently of one another—didn't work. Despite mutual artistic respect, "it was a collaboration," said Foreman, "in which none of the collaborators discussed things with the others."[22] Many critics and much of the audience voted early with their feet. There were those, however, who strongly defended Acker's often obscene text as a necessary feminist repossession of the language of desire.[23]

Foreman had considerably more popular success with some later forays into opera. At the American Repertory Theatre in Cambridge in 1988 he staged Philip Glass's musical setting of *The Fall of the House of Usher.* Attracted by the tension between the tale's feverish emotions and the cool surface of Glass's music, Foreman found many imagistic affinities between his signature style and Poe's imagination: spinning mirrors, descending chandeliers, the front drop fringed with skulls, etc. And in 1991 Foreman returned to France to baptize a new opera house in Lille with a production of *Don Giovanni,* which was well received. Ironically, in light of the *Fledermaus* furor, the *Time* magazine music critic held the *Don Giovanni* production up as an antidote to the directorial excesses that had recently plagued opera (by Peter Sellars et al.). Foreman, he asserted, had taken the middle path between slavish traditionalism and directorial hubris in a respectful, but not at all worshipful, version of Mozart's masterpiece.[24]

Musical Interludes

Early in his career, Foreman decided, as a counterweight to his admittedly difficult serious plays, to devote part of his energies to more accessible work. The result has been a species of musical theater closer to commercial theater than his Ontological-Hysteric productions. But not *too* close: for he did not abandon his perennial concerns; he merely softened them. Long attracted to the vitality of popular forms (which he continually uses in his O-H work), Foreman found himself engaged in co-creating a series of musical comedies with a difference. Without relinquishing his core ambition to create provocative avant-garde work, he simultaneously hoped to create musical theater which, "while accessible, created an atmosphere of poetic mystery."[25]

His musical collaborator for over twenty years has been Stanley Silverman. The two met early in Foreman's career (through their wives, who had been classmates) right after the production of *Angelface,* a copy of which Foreman sent Silverman. To his surprise, Silverman asked him to write the text for an opera he had promised the Tanglewood Festival. The result, *Elephant Steps,* was hardly a traditional libretto: the text focused on a young man's initiation, but it was narratively discontinuous, replete with grotesque images of elephant angels, and stylistically indebted to German Expressionism. Fully a third of *Elephant Steps* consisted of abstract phrases that Silverman set to atonal music. In the rest of the piece, the music evoked more popular forms: 1930s swing, romantic ballads, and rock and roll. It took two years for the piece to reach New York City, where it made an undeniable impact. Richard Kostelanetz hailed it as "something new and important in the history of American theater—a truly *contemporary* opera. . . . As a theatrical experience *Elephant Steps* is stupendous, multisensory, original, diffuse, overwhelming, faintly frightening, and always surprising."[26] For the text, Foreman received the first of his many *Village Voice* Obies.

In 1972 Foreman and Silverman had their greatest popular success: *Dr. Selavy's Magic Theatre.* Originally presented at the Lenox Arts Center in Massachusetts, the piece went on to a long run Off Broadway. After subsequent staging in places like Cleveland, Ohio, and Oxford, England, it was revived in New York in 1984. Silverman's songs had originally been written for a production *manqué* of *The Satyricon* at Canada's Stratford Festival, and Foreman arranged them in a new context inspired by his responses to the music. The punning title ("c'est la vie") was borrowed from Marcel Duchamp; the piece itself—in which a mental patient pursues an elusive cure—can be seen either as a course of psychiatric treatment or a spiritual quest. The mood is one of delirium, but a delirium moderated by Silverman's pastiche score, which evokes a musical revue, "a twenties revue," the *New York Times* suggested, "gone beserk in the seventies."[27] When he revived the piece with trepidation in 1984, Foreman stated that "this production is more intricate and faceted. . . . It's like watching a spinning top. Originally it had a goofy '60s feeling. Now it's end-of-the-world '80s."[28]

The collaboration between Foreman and Silverman has survived the decades because each has maintained respect for the other's independence. "[In] my collaboration with Stanley," Foreman has stated, "we stay out of the other's way. After we've discussed a project, I write the text, send it to him, he's free to do whatever he wants with it, he sets it, hands it back to me, I'm free to do whatever I want with it on the stage."[29] From *Elephant Steps* in 1968 through *Love & Science* in 1990, Foreman and Silverman produced eight diverse music-theater works, which walk the line between avant-garde seriousness and popular accessibility. One of their most popular, *Hotel for Criminals* (1974–75), was inspired by Louis Feuillade's early silent film serials devoted to the arch-criminal Fantomas. The piece presents a subterranean world that deepens into surreal nightmare. Silverman's score skillfully juxtaposes French *chansons* with high Baroque, Lullyesque music so that the nightmare world always contains an element of the parodic and the absurd.

In 1984 Foreman and Silverman again collaborated on a piece (opening in Massachusetts, *not* initially staged by Foreman) that surfaced two years later in New York, now under Foreman's direction: *Africanus Instructus*. In the spirit of earlier Ontological-Hysteric pieces on the theme of the disequilibrium of travel, *Africanus Instructus* centers on an adventurous young innocent who, this time around, is a proper young Victorian thing (replete with governess) visiting the Dark Continent at the end of the nineteenth century. She is indeed named Rhoda but she was not played by Kate Manheim. As in *Egyptology,* social themes inevitably arise from the racial iconography (one of the hunters is white, another black), but they are secondary to the exploration of the dark continent of consciousness, particularly the great divide between instinct and civilization. At the very same time that white imperialists

were exploring Africa, Foreman notes in his introduction to the published version, Freud and Jung were exploring the dark unconscious.[30]

Silverman's prolific score—more than forty songs and recitatives—
serves as counterpoint to the perennial Foremanesque concerns. Tran-
scending mere pastiche, the music modulates between popular and
high art forms, now music hall, now Romantic rhapsody, now the hint
of a Puccini aria, etc. But, though Foreman and Silverman were pleased
with their collaboration, the American critical consensus was that
"somewhere in the African wilderness . . . Mr. Foreman has misplaced
his sense of playfulness."[31] In Europe, however, the work was well
received.

Less Is More

In 1987 Foreman produced a play in collaboration with New
York University's Tisch School of the Arts with the tantalizing title *Film
Is Evil: Radio Is Good.* The play is set in a bizarre Foremanized version
of a small radio station, where the station personnel strive to withstand
a takeover by cinéastes who insist that only that which can be filmed—
"the concrete physical world . . . is therefore the real world."[32] Film is,
therefore, *the* art of Truth. Radio's defenders, while admitting film's sensuous power, insist that this power is hollow; film is a trick of perception, an illusion. The WORD is truth and radio is its prophet. The balance (or is it a struggle?) between image and word is, of course, at the
heart of Foreman's work, something he guiltily recognizes: "Visuality is
evil, and yet I'm a great indulger in visuality. It's a paradox, but I'm im-
mersed in that evil."[33] And so he felt a need to simplify and redirect his
creative energies back to the word, the text. From the mid-eighties
on—with certain crucial exceptions—his work has been more intimate,
softer, reduced in scale. In the previous decade, he had worked with
about a dozen performers, many of whom functioned as a chorus.
Henceforth, three to five performers become the norm.

Part of this response was the result of his collaborations with the
Wooster Group in their home space, the Performing Garage. Like
Joseph Papp, the Wooster Group provided Foreman with a place to
work after the closing of his Broadway loft. From 1985 to 1990 Foreman
presented four plays at the Garage: two with the Wooster Group ensemble, *Miss Universal Happiness* (1985) and *Symphony of Rats* (1988);
and two without, *The Cure* (1986) and *Lava* (1989). Foreman acknowledges that the visceral acting style of the Wooster Group—so different
from the ironic deadpan he usually coaxed from his performers—

clearly moved *Miss Universal Happiness* in hyperkinetic directions: "[The] actors from the Wooster Group . . . didn't hesitate to expand on my suggestions to get louder and faster. . . . They threw themselves into it with a vengeance that moved the performance faster and faster, and I indulged myself by asking for more violent physical activity than I ever had in the past."[34]

Set in the cartoon milieu of a Third World Banana Republic in crisis, *Miss Universal Happiness* allows the subterranean politics of plays like *Egyptology* and *Africanus Instructus* to erupt like lava onto the surface of the play. The eponymous heroine, played by Wooster stalwart Kate Valk, is a colonialist, Rhoda, who carries the torch of "liberty"—the American Dream—to the exploited; the latter, however, unable to attack their real enemies, turn on each other. To capture this violent mood of manic nightmare, the production presented a verbal, visual, and aural sensory overload, with many scenes achieving a painful saturation level of noise and action.[35]

After *Miss Universal Happiness,* Foreman became convinced that he could go no further in the direction of Dionysian frenzy and determined to explore, in his words, "alternative dynamics." For his next piece at the Garage, he used, not its main space, but a small room that usually served the Wooster Group as a dressing room. *The Cure* had only three characters, who carried the first names of the actors who originally played them. The actors wore body mikes that allowed them to deliver their lines at a volume little above a whisper. The customary Foremanesque scenographic provocations—buzzers, blinding lights, etc.—were conspicuous by their absence. The play aimed for a contemplative, inward, hushed sensibility; to this end, Foreman rehearsed his cast with uncharacteristic Stanislavskian exercises to release more internalized, personal performances. In his earlier Ontological-Hysteric pieces, Foreman had scrutinized the recalcitrance of the external world when confronted by the forces of consciousness and perception; now his aim was "to make a play that existed in the internal world itself, with no concern about projecting that way of being into the world outside the self."[36] Essentially, the model he followed was that offered by Strindberg for his chamber plays: "intimate in form; . . . few characters; vast perspectives; freely imaginative but built on observations, experiences carefully studied; simple, but not too simple."[37]

As he has grown older, Foreman's sensibility has turned even more autumnal, as evidenced in chamber plays such as *What Did He See?* (1988) and *The Mind King* (1992). But his perennial aesthetic restlessness always resurfaces; more recently, he has turned again to overtly theatrical plays, though on a lesser scale than before, with the casts remaining small and the running time usually a little over one hour.

Fortunately, Foreman's creative circumstances have been eased in the 1990s by the end (let us hope) of the Ontological-Hysteric Theater's peripatetic existence. In 1991 Foreman took over the historic upstairs theater at St. Mark's Church on Second Avenue in New York's East Village (which in the 1960s had housed one of the first Off-Off-Broadway theaters, Theatre Genesis, which produced Sam Shepard's beginning plays). This space, which has been home to Foreman's most recent work, has helped shape his scenography, just as his earlier work was influenced by his Broadway space; while the latter was deep and narrow, hence accenting the aperspectivity and distancing, the St. Mark's space is wide and shallow, providing the expanse of a large Cornellian box yearning to be filled with objects.

After the relative scenic simplicity of the inaugural play in the new space, *The Mind King* ("a dimly illuminated room, walls painted gray"), the St. Mark's plays have increased their visual component. *Samuel's Major Problems* (1993) is set in a littered room in which a party has taken place; balloons, streamers, loose paper, confetti, are scattered about; pictures of skulls on the wall replicate some real skulls mixed in with other strange objects on the shelves. In *My Head Was a Sledgehammer* (1994) the set is a kind of classroom with blackboards and tall bookcases and tiny flags hung around; roses creep up the walls. It could be a lecture room, or a studio, or a gymnasium—or all of them at once. *I've Got the Shakes* (1995) repeats the image of (doomed?) festivities: balloons and streamers again festoon the space, white flowers cascade near the ceiling. In *Benita Canova* (1997), inspired by the erotic paintings of Balthus, the set is a sumptuous nineteenth-century *fin-de-siècle* sexual preserve: a bordello or a back room for liaisons, lushly carpeted and draped in black, gold, crimson, furnished with an assemblage of couches, baby dolls, and garden gnomes.

Characters and narratives remain typically elusive, but themes are perhaps more accessible than before. With age, Foreman has turned toward themes of death (the memento mori of skulls in the imagery) and self-assessment. The party's over and was it any good? *Samuel's Major Problems* is a most somber contemplation of death's approach: at the end of a long party, Samuel is chased by a young woman who may be the Angel of Death and a doctor who may be Death also. In *My Head Was a Sledgehammer*, a professor fails to pass on his legacy. Foreman's "plot" summary on the original program charts the essential theme (Foreman is usually his own best critic): "Through techniques that seem weird and bizarre, the Professor teaches a powerful poetic method . . . his students try to echo. . . . But imitating God (he who 'knows'), the Professor falls from love's kingdom and re-emerges as an old man on leaden feet. . . . His students turn away from him towards their own illusionary dreams of happiness."

This theme of painful artistic self-scrutiny is perhaps most definitively articulated in *Pearls for Pigs* (1997), which, though produced by the Hartford Stage and International Production Associates, remains very much an Ontological-Hysteric play. (The production toured nationally and internationally before being presented at the Tribeca Performing Arts Center in New York.) Simultaneously a Pierrot show, a farce, a comic Grand Guignol, a Pirandellian mystification, and a manifesto, *Pearls for Pigs* nakedly offers us the author's intellectual self-portrait in the character of the Maestro, who ventures "to sum up approximately thirty years of theatrical adventurism." To Pierrot he explains his aesthetic: "Get this straight! You are not the main character. I am not the main character. The world itself, is the main character."[38] Since the world is ultimately unknowable, all one can do is "disguise reality with a mere facade built on totally imaginary foundations": hence, theater where "I am safe inside the walls of my private kingdom." But the safety is precarious: "Light from the stage blinds me . . . and I am paralyzed in front of the not yet risen curtain." Is it worth it? To expose oneself to the agonies of "mind attack" for an audience that wants *its* mind unviolated? Is it not indeed casting pearls before swine? The maestro "ruins" his play by ordering his characters off stage. But, bemused by his own perversity, he calls them all back, commanding: "Ask nothing. Mysteries should never be solved. Dance to the music of that commandment, please. Everybody." As they all whirl the beat goes on.

Richard Foreman will persevere. Like Beckett's Unnamable, he can't go on, he'll go on. As he confesses in *Permanent Brain Damage* (1996), "an angry man, disillusioned and weary of life, nevertheless sustains the blind faith that something amazing and beautiful will emerge from the chaos surrounding him." In his most accessible book, *Unbalancing Acts* (1992), Foreman noted that "essentially, an artist does one thing throughout his career."[39] Indeed, Foreman has continually recycled scenic images (strings, blinding lights, etc.), titles *(My Head Was a Sledgehammer, I've Got the Shakes),* and intellectual concerns. The steadfastness of this vision remains unique and indispensable, and, yes, adversarial: he maintains that when conservative critics attack contemporary art because they fear that it will undermine western culture, they have a valid fear. Not now, but "in the long run." Until then, Foreman will keep going—thirty years and counting—by continuing to create serious art in a culture hostile to serious art, "listen[ing] carefully . . . to hear the approaching footsteps" of amazing and beautiful order.

Notes

1. Randy Gener, "Yut Ta Ha," *Village Voice,* June 1, 1993, 94.

2. "Foundations for a Theatre," in *Unbalancing Acts: Foundations for a Theatre* (New York: Pantheon Books, 1992), 31. See also Foreman's declaration of the need for "exciting, revolutionary . . . counterculture theatre" in a *Village Voice* symposium, "What's Ahead for Off-Broadway?" May 29, 1990, 69.

3. "From the Beginning," in *Unbalancing Acts,* 74.

4. Ibid., 73.

5. Ibid., 75.

6. See the essays on "Both Halves of Richard Foreman" by David Savran and Arthur Bartow, *American Theatre,* July / August 1987, 14–21, 49.

7. "From the Beginning," 71.

8. "Visual Composition, Mostly," in *Unbalancing Acts,* 71.

9. "Rhoda in Potatoland," in *Richard Foreman: Plays and Manifestos,* ed. Kate Davy (New York: New York Univ. Press, 1976).

10. Richard Schechner, *Performing Arts Journal* 14 (1981): 50.

11. See Kate Davy's essay and Richard Schechner's review of *Rhoda in Potatoland* in this collection.

12. "From the Beginning," 74.

13. "Rhoda in Potatoland," 216.

14. Marc Robinson, *The Other American Drama* (New York: Cambridge Univ. Press, 1994), 167.

15. "Rhoda in Potatoland," 220.

16. Timothy na Gopaleen, "Richard Foreman: Down and Up in Paris," *Village Voice,* December 27, 1983, 107.

17. "Egyptology," in *Reverberation Machines: The Later Plays and Essays* (Barrytown, N.Y.: Station Hill Press, 1985), 165.

18. Clive Barnes, *New York Times,* May 3, 1976, 42.

19. The Venus in question is the Hottentot Venus, whose voluminous behind made her a sensation in early-nineteenth-century freak shows. Suzan-Lori Parks, a talented African-American playwright, mines this historical event for its delineation of racial and sexual exploitation. Some felt that Parks's points were *too* obvious, that her sensibility did not mesh with Foreman's, and that their collaboration was a shotgun wedding (Ben Brantley, *New York Times,* May 3, 1996, C3). Others agreed with John Lahr that the play was "a brilliant

meditation on the ambiguity of race, history, the colonized imagination, sexuality, and theatrical storytelling itself" (*New Yorker,* May 6, 1996, 98).

20. Frank Rich, *New York Times,* June 7, 1981, 25.

21. Thomas Quinn Curtis, *Variety,* March 16, 1983, 191.

22. Michael Feingold, "An Ungrammar of Ornament," *Village Voice,* December 17, 1985, 120.

23. See, for example, Erika Munk, "Semper in Absentes," *Village Voice,* December 19, 1985, 117.

24. Michael Walsh, *Time,* June 17, 1991, 48.

25. "Brightly Lit Crystal Chandeliers," in Richard Foreman, *Love & Science: Selected Music Theatre Texts* (New York: Theatre Communications Group, 1991), xii.

26. Liner notes to Columbia recording of *Elephant Steps,* 1974.

27. Stephen Rubin, "The Magic Theatre of Dr . . . Who?" *New York Times,* Arts section. December 17, 1972, 21, 40.

28. Mel Gussow, "Dr. Selavy," *New York Times,* January 27, 1984, C3.

29. Michael Feingold, "Can These People Save Off-Broadway?" *Village Voice,* January 13, 1975, 80.

30. *Love & Science,* xiv.

31. Mel Gussow, *New York Times,* January 26, 1986, C43.

32. "Film Is Evil: Radio Is Good," in *Unbalancing Acts,* 170.

33. Ibid., 147.

34. Ibid., 97.

35. Foreman's second collaboration with the Wooster Group, *Symphony of Rats* (1988), was less frenetic but still bizarre. Its plot: in the future, with the Earth's natural resources depleted and food by necessity artificially manufactured, a president of the United States (played by Ron Vawter, originally intended for the unavailable Willem DaFoe) experiences hallucinations that appear to be messages from somewhere else in the universe. He is transported to some mysterious planet where he encounters, among other disturbing things, strange rat-like figures with video monitors for faces. The play's broader theme is that the human challenge is not only to escape the false messages of our corrupt culture but also to make ourselves receptive to "other" forces and energies that can spiritually release us. It is not salvation, however, but the noise of destruction that is represented by the play's final image: an electronic hum escalates to the obliterating roar of an approaching air squadron as the president looks to the sky fearfully in the enveloping darkness.

36. "The Cure," *Unbalancing Acts,* 107.

37. August Strindberg, *The Chamber Plays* (New York: E. P. Dutton, 1962), vii.

38. Quotations from author's original typescript.

39. "From the Beginning," in *Unbalancing Acts,* 85.

I Overviews

Kate Davy

Kate Manheim as Foreman's Rhoda

Kate Manheim spent most of her life in France, coming to the United States in the fall of 1969 and to New York City in March, 1970. Because she speaks two languages without an accent, one of her first jobs in this country was at the Berlitz School of Languages, where she taught private lessons in both French and English. In the fall of 1971, shortly after she began working at the Anthology Film Archives, she met playwright / director Richard Foreman. Foreman was already in rehearsal for his Ontological-Hysteric Theater's fourth production, titled *HcOhTiEnLa (or) Hotel China.* Since one of the actresses had dropped out of the play, he went to the Anthology looking for a volunteer replacement. After walking back and forth several times through the office where Manheim was working, Foreman finally approached her and asked shyly, "Would you mind being an angel in my play?" She replied simply, "No, I wouldn't mind."

Although she had never heard of Richard Foreman or his plays, there was something that attracted Manheim to him immediately. She describes it in terms of the peculiar way he opened and closed doors. He did not just close the door behind him but instead, after walking into the room, he turned, faced the door and pulled it toward him, closing it before turning to continue through the room. Watching a rehearsal that evening, Manheim noticed that all the performers were closing doors in the same, somewhat disjointed, manner.

In the last five years, Manheim has performed in nine of Foreman's twelve Ontological-Hysteric Theater productions. (Although Foreman has written thirty-nine plays since 1967 and directed and designed twenty-one of them, only twelve were produced under the title of "On-

Originally published in *TDR* 20 (September 1976): 37–49.

tological-Hysteric Theater.") A character named "Rhoda" appears in ten of the twelve plays and Manheim has played this character exclusively in seven of them beginning with *Sophia = (Wisdom) Part 3: The Cliffs,* 1972–73. In October 1976 she will play Rhoda again in Foreman's play *Livre des Splendeurs,* which will be performed in French for the Festival d'Automne in Paris. Manheim is translating the text for this production.

While Manheim had never appeared onstage before *Hotel China,* she did have two previous encounters with actor training. In Paris, she attended the Ecole Charles Dullin at the Théâtre National Populaire, where she was required to perform scenes from classical French plays. "I wanted to be an actress, so I went to school but found it very uninteresting, detached from any reality, and it scared me to death." She quit after two months and decided that if she were going to be an actress, "it would have to happen in a different way." Her second experience was more profitable in terms of training.

Ruth Maleczech, one of the original and current members of Mabou Mines, was living and working in Paris in 1968. Beginning in the spring of 1969, Manheim and Maleczech got together on a one-to-one basis to, as Maleczech put it, "fool around" with various acting exercises. "Ruth taught me about Stanislavsky and certain tricks like thinking about something else when saying a line." During the six-month period they worked together, doing improvisations as well as scenes from plays, Manheim learned about exteriorizing emotions and ridding herself of inhibitions. One experiment involved performing a scene first in English and then in French. Much to her surprise, Manheim discovered that because her sensitivities were different in each language, the emotions of the scene were not the same. "I discovered that I'm a very different person in French than in English."

The time she remembers most clearly, and that was the most useful in terms of her current work, was the session when Maleczech asked her to sing a song. Manheim explains that singing has always been a tremendous block for her and, "I just couldn't do it—I just couldn't get it out." It was an extremely traumatic experience, but after almost four hours, trying numerous approaches, Maleczech finally got her to sing an entire song. Manheim remembers, "Afterwards, I cried a lot." In Foreman's most recent piece, *Rhoda in Potatoland (Her Fall-Starts),* 1975–76, Manheim had to sing a few simple lines of a song, and she is convinced that, had it not been for that experience with Maleczech, she would not have been able to do it.

In terms of performing, there are two common misconceptions regarding Foreman's work. One is that there is no "acting" in his theater pieces, and the other is that the character "Rhoda" did not appear in the plays until Manheim began playing her. While it is true that Foreman and Manheim began living together several months before Manheim first played Rhoda in 1972, the character appeared in Foreman's first production, *Angelface,* in 1968, as well as several produced and unproduced plays prior to *Sophia = (Wisdom) Part 3: The Cliffs.* Perhaps, one of the reasons it is assumed that there is no acting in Foreman's theater is the fact that he usually does not use people with acting or performing backgrounds. There are no auditions for casting a Foreman production. Occasionally he will ask someone to be in a play but generally he simply accepts volunteers.

Another reason for this "non-acting" assumption is based on the unusual nature of Foreman's characters. Although many of the same characters (Rhoda, Ben, Sophia, Max, etc.) reappear in all of the texts, it is extremely difficult to identify with these characters, regardless of how familiar the reader/spectator becomes with them. This is because Foreman's characters are not presented in definite, recognizable situations and they do not, except momentarily, have feelings, needs, goals, or ambitions. They function more as self-enclosed units than as characters responding or reacting to situations of conflict or encounter. Occasionally, they even refer to themselves in the third person, thereby distancing the viewer and reducing the possibility of emotional involvement with the character. It can be said that while playing such a character involves acting, it is a particular kind of acting—certainly different from what is understood as "method" acting. The distinctive kind of performing involved in a Foreman production can, perhaps, be described and elucidated by examining a particular performer—Kate Manheim—and her personal approach to, and playing of, a specific character—Rhoda.

Many people tend to interpret the plays in terms of Foreman's and Manheim's personal lives. This is alright as far as Manheim is concerned, but she does not think about it. She does not see herself as Rhoda, nor Foreman, necessarily, as Max or Ben.

> It's me to the extent that I live with Richard—and any part that he writes has something to do with me and him, whether it's Max or any of the other women in the plays. Since I'm part of his life, everything has a little bit to do with me, as well as other things in Richard's world. But I don't identify with Rhoda that much.

To presume that Rhoda is Kate Manheim and does not exist separately from her is not totally unfounded considering the essentially limited nature of the character. However, Manheim sees the plays themselves, and therefore Rhoda, as basically foreign to her reality. Regarding her relationship to the character in *Rhoda in Potatoland,* Manheim explains, "It's a character, and when I play that character I use things that are somewhere in myself. The way I portray the character isn't completely me although I use things that I have and that are a part of me—things that I can do. I'm not in 'potatoland'—she is. It's a character who has to face certain adventures—but they are not mine."

Manheim readily admits that in approaching the role it is not possible to make a list of characteristics unique to Rhoda's personality. Since Rhoda has no past or future, the character exists only in the present, and Manheim makes no attempt to impose a history on Rhoda or imbue her with personal traits or idiosyncracies separate from her own presence on the stage. Hence, the processes, methods, or techniques most often associated with "creating" a character do not apply, and consequently, Manheim's approach to playing the role proceeds in an entirely different direction, beginning with the first weeks of rehearsal.

Rehearsals are rather austere and can be somewhat tedious for the performers since the most distinctive feature of the rehearsal process involves Foreman's rigorous attention to, and accumulation of, the most minute details of staging. He controls every aspect of the work—the performers neither improvise nor initiate activity. He merely expects them to write down all of their actions, positions, and pauses; memorize them; and carry them out as precisely as possible. He does not explain "why" an activity must be carried out in one way rather than another, and the performers seldom ask questions or suggest solutions. Because his staging process is closer to that of an artist working in private (writing, painting, or sculpting) than that of a theater director, often what seems to be a specific instruction to a performer is actually Foreman thinking out loud or talking to himself.

Two and one half years ago, during rehearsals for *Pain(T)* and *Vertical Mobility,* Manheim said:

> I get the impression that usually in rehearsals—not that I've worked in any other plays—one works in different ways, getting different directions from the director and then working on things. With Richard you don't get that experience because he has one idea, and when you try it—let's say you can't do it at that moment—he'll change it to something you can do immediately

instead of working on the thing and learning "how" to do it. Sometimes I get the idea that I'm not really given a chance.

Foreman is still generally unconcerned with a virtuosity of acting, but now he and Manheim do rehearse certain sections on the weekends, without the other performers present. While he does not indicate how she should feel or what she should think when delivering a line, he does tell her how to say it. For example, he told her to deliver a particular speech from *Potatoland* about a boat so that it sounded something like a lecture, and then she worked on it until they both felt it was "right."

Within the overall and very precise framework Foreman gives her to work in, Manheim feels she has the freedom to behave or "act" in her own personal way.

> At the beginning of the rehearsal period I have to memorize all the precise actions, and that takes a number of weeks. I work so that all of the outline is very precise and everything Richard wants is in there and then, within that, I find I have a great freedom to "act" in a certain way.

Because in Foreman's productions objects and set pieces are often given the same status as live performers, many people contend that Foreman uses or treats his performers like objects. Manheim commented:

> I think it's very unfair to say he treats people like objects because I feel more free—much more myself and much more a person—in Richard's plays than in any other place in my life. The extreme preciseness of the outline, it's only an outline he gives people—it doesn't have anything to do with their souls— within that outline somehow, that frame, people are much stronger and much more themselves.

One concrete, if minor, way to express one aspect of what Manheim means by "more themselves" is the fact that each performer chooses what he or she will wear in the performances—usually their everyday street clothes.

Thus, in the first several weeks of rehearsal, Manheim concentrates solely on memorizing and carrying out the physical tasks precisely. She explains that, like Foreman, she works against ambiguity. "Just as Richard tries to make each moment of the play clear, I work to make each thing that I do completely clear to the audience." It is not until about three weeks before each play opens that she begins to have a consciousness of this particular "Rhoda"—what she is doing and her place in the piece as a whole. More importantly, she does not worry about it at all until this point in rehearsal. "I wait until the run-throughs, until I see where the whole thing is going and then I begin to work with that." She never concerns herself with what the play is "about." Furthermore, when asked if it mattered whether or not she understood the plays, she replied, "No, it doesn't matter and I don't really understand them, in his terms. I do have an understanding, but it's more intuitive—I couldn't explain the plays in the same way Richard would."

The idea of the plays as removed from Manheim personally is basic to understanding her approach. The world of each play is Foreman's, and she does not attempt to make it hers or interfere with his vision of the work. Instead, she tries to fit herself, not the character, into a world of objects, situations, and images that are outside of her reality. She admits that she does not know what Foreman means by "potatoland" and that the entire notion is very far from her. Consequently, she works at developing ways to make herself feel comfortable in a world that is not hers. The result of this process is the character of Rhoda. Rather than forcing her own thinking and feelings on the work, she lets the character come through her. While Rhoda and Manheim cannot be equated, they cannot be completely separated either—since Rhoda is not a fully developed character, Manheim cannot hide behind an interpretation of the role. Ultimately, Rhoda is an expression of Manheim—the end result of her singular work on the character. This work relates more to the use of personal techniques and particular states of mind than to the development of a specific character.

If her own individual work begins late in the rehearsal period, what precisely does it involve? During the initial run-throughs of *Rhoda in Potatoland,* Manheim was very unhappy because, although she was onstage constantly, "I felt that I didn't have anything 'to do' in this play." This feeling was the result of the fact that the pacing in *Potatoland* was quite a bit faster than that of previous productions. She felt everything was going past her so quickly that the character was invisible. Hence, she focused her attention on learning how to do everything even faster than it had to be done, in order to give herself a few moments to think. Gradually, she began to master her physical activities in the context of the pacing so she could begin concentrated work on the character. It was not until the last two weeks of rehearsal that she began to "connect" with the piece as a whole and her attitude toward the work changed. She explains, "It wasn't until the last two weeks that the interesting work began for me."

Manheim has stated that over the years many audience members have commented especially on (1) her "presence," (2) "the Rhoda look" or the use of her eyes, and (3) her energy level. She has developed and uses specific techniques to achieve these physical states or qualities. These techniques can be separated into two categories—the mind, or work relating to concentration, and the body, or specific physical work. In addition, these techniques are closely related to generating various "states of mind" during the performance.

In terms of focus, Manheim employs an entire continuum of thought ranging from total moment-to-moment concentration to a complete absence of concentration or allowing the mind to wander. While at times she focuses her attention on the present, concentrating exclusively on the moment-to-moment progression of what she is doing, there are other times when she thinks about the future, enumerating in her mind the activities to follow. She explains that when she is onstage, she finds time when she is not saying anything to recapitulate "everything that will follow for the next few minutes until the moment I know I'll have another little break where nobody will notice that I am thinking about what I am going to do next." Finally, she concedes that quite often her mind simply wanders and she thinks about things completely unrelated to the play and her performance. However, she makes no attempt to control this phenomenon—instead, she allows it to happen, turning it to her advantage for a particular effect. She explains, "I try to keep myself off-balance in a certain way to keep myself in a state where I'm always surprised by what I'm doing or what's happening." Two years ago, she described this impulse in terms of a specific technique:

> I blank out my mind until the last possible second when something has to be done, and then there is always the possibility that I'm going to forget it, but I don't somehow. I keep myself at a limit where it's sort of dangerous and exciting and I suppose it shows.

Although she no longer attempts to blank out her mind, the technique has been assimilated and the procedure is more unconscious. Allowing her mind to wander serves somewhat the same purpose.

In relation to what many spectators have labelled "presence," Manheim works with her body in certain ways to "make myself feel more intense when performing." For example, while simply walking from one point to another, she will "tighten up" muscles not used for carrying out that activity, perhaps consciously contracting the muscles in her stomach or arms. In general, she attempts to push her body into uncomfortable positions. She explains, "I always work against whatever comes naturally to the body." Often, when she is merely standing motionless onstage, one might notice that her foot is turned in a somewhat awkward direction or an arm might be slightly twisted into an uncomfortable position. In fact, if her body gets accustomed to such a position through repetition, she will make the position a little more extreme, thereby pressing the body in order to recapture the original feeling of discomfort. When the tasks and positions Foreman gives her are difficult and uncomfortable, which they frequently are, she works with, rather than against, them, driving the body toward extremes of activity.

In her work, Manheim sees a relationship to the Russian ballet dancer and choreographer Nijinsky. "I don't want to seem presumptuous, but there is something about our energies that is similar." Indeed, she sees her physical work on a continuum between free, natural movement and utter stasis. "There is a very strange fight going on in myself. I try to work at a limit between what the body would be expected to do normally at one end and a point where the movement is so unnatural that the body can't move at all." More than anything else, onstage she is aware of her body. She does not put herself in a certain psychological state, allowing the body to follow, but rather, "if I'm doing a certain thing with my body, I'll feel the right way."

In her mind, she considers individual gestures and movements as consisting of several independent units. "I think of the movement I'm doing in terms of breaking it up." Although each gesture is carried out swiftly and appears smooth and connected as a whole, she explains that, "I think of each gesture as all of the fractions of movement that make up the general movement." While her movements do not come across as a series of short, broken, or choppy component parts, the fact that she imagines them in a manner similar to Duchamp's painting *Nude Descending a Staircase* lends a certain quality, perhaps tension, to her general movement pattern.

Another important feature contributing to her onstage "presence" involves the extensive work she has done with her eyes. Because Foreman frequently positions his performers so that they are gazing steadily at, toward, or beyond the spectators, Manheim developed a way of looking out in keeping with her general tendency toward maintaining physical discomfort. Especially in her earlier performances, she concentrated on not blinking her eyes at all, until she could do this, while looking blankly out at the audience, for long periods of time. Although in the last couple of years she has abandoned this conscious effort and concentrates instead on relaxing her eyes, the highly focused quality of her gaze is still present. Since *Vertical Mobility* (1974), she has developed an additional way of working with her eyes that she uses occasionally during each performance. It involves moving the eyes in succession back and forth very rapidly, producing an almost "manic" appearance or impression. The continuous intensity associated with "the Rhoda look" relates more to a general energy level than to tension in the sense of strain, stress or anxiety. This energy level, while not identical to tension, is related to a kind of "nervousness" that she consciously promotes or fosters before each performance.

While she does not spend time on physical or vocal exercises, she does put an hour aside before each performance, until the last day of the run, to prepare herself. This preparation is related to her desire to maintain a constant state of surprise. Although she does not pretend to be doing the play for the first time when she is performing, she spends the hour beforehand reading through the entire script as though she had never done the play before. Because there is a quality she does not want to lose during a long run, she explains that each night, "I sort of cultivate my nervousness about the whole business." Hence, she is concerned exclusively with preparing herself psychologically.

While onstage, Manheim is aware of a certain "ecstatic" state that she does not usually experience in her everyday life. She describes this as a general mental condition relating to a kind of vitalized or highly animated existence. Before each performance, she explains, "I put myself in a state of mind where I'm the most 'alive' possible." She considers this feeling of "aliveness" as an essential dimension not only of her work as a performer but of her life as well.

> The plays—being onstage for that time and the rehearsal period which is the working out of it—are a training in making myself the most alive possible. It all has to do with everyday life but it's much more concentrated. I need the feeling, just to continue living, of concentrating that "aliveness" in certain movements of my life. One does this, for example, in sex, also.

She finds that this "concentrated aliveness" is not generally present in her everyday life and, in fact, could not be because, "If I was that alive all the time, I'd be dead in twenty-four hours." By being the "most alive possible" she means being the most focused within herself, which she associates with an extremely high energy level. When asked if this related to being the most "present" onstage, she replied, "It's not the most present but rather the most aware of being present."

This awareness relates to the amount of intellectual thought occurring, on the part of the performer, during a performance. Manheim is continuously aware of the process of performing while she is onstage.

> I'm not interested in letting myself get carried away in a role and becoming so absorbed in the character I'm playing that I forget and lose myself. I'm not interested in it all coming so naturally that when I say the lines it's like the person [character] saying them.

While she wishes to remain detached from the role, maintaining a certain distance, she does not attempt to comment on the character while she is playing it. She works neither in terms of interpretation nor motivation while developing and executing her performance. When asked if Foreman's "cue system" (which involves periodic live sounds, such as buzzers, combined with tape recorded music and noises) provided the primary motivation for getting from one activity to another, she replied, "Yes. What else?" Furthermore, instead of describing any specific devices she uses for delivering her lines, she explains that it is merely, "a matter of organizing how to say the line." Perhaps by examining her thought process during a particular scene the precise ways in which she works can be demonstrated.

In *Rhoda in Potatoland,* Manheim performed a sequence on top of a table that was slanted toward the audience in such a way that she had to hang onto the sides to keep from sliding off. The scene began as Manheim walked onto the stage with her hand to her chin, and paced back and forth directly in front of the spectators, looking as though she were rather skeptically "sizing them all up." She explains that while she was doing this, "I would think to myself, 'Oh, you're still there.'" This thought would come across as more or less hostile depending on how she felt about a particular audience on a particular night. In other words, at that moment, Kate Manheim, as performer, was consciously relating to and responding to the spectators' presence. Next, as loud music began to play, she slid onto the table and a large, black, rubber galosh placed on one of her feet. She acted terribly frightened of it, raising and holding her leg up off the table, as if she were trying to stay as far away from the boot as possible. A potato was then brought onstage and placed on her stomach. From the moment she saw the potato, she became increasingly more hysterical, gyrating her body back and forth wildly, as much as she could while holding onto the table. Although she considered her acting in this scene as representing a kind of hysterical fit, she made no attempt to recreate such an emotional state inside herself. She explains that she was creating "just the appearance of such a state"—going through the "motions" of hysteria rather than living through such an experience emotionally. "I didn't take those potatoes that seriously, so I couldn't have actually been living it."

In addition, when working on the sequence, she did not think in terms of making the hysteria "believable" in the sense of convincing the spectators that the character was really experiencing such an ordeal. She was not trying to manipulate the spectators' emotions, coercing them into identifying with Rhoda, thereby feeling anxious or sorry for her. When asked how she went about working on the scene, she began by explaining how it was carefully staged and then, "I got used to the scene and I knew that each thing was a cue [like seeing the potato] and I react according to plan." This, of course, does not mean that because she reacts automatically, her acting style is not as good as those using different approaches. On the contrary, her energy level and work with the body makes her total performance quite skillful and impressive. The fact that she is not required to become psychologically or emotionally involved with either the character or a scene merely gives her the opportunity to concentrate on other work—the physical tasks in the hysteria scene, particularly, were extremely difficult and strenuous.

Concerning the amount of intellectual thought involved in Manheim's performing, she explains that in rehearsals she works primarily intuitively until she, and Foreman, both agree that a certain activity or action is "right." Then, "once we've got it right—that is, I feel O.K. doing it and he thinks it's right—when performing it I think about those times when it was right—I think about that more than anything." In addition to thinking about how it felt when she was doing it correctly, immediately afterwards she frequently thinks about whether it went well or badly. While performing, she thinks about the "way" she is performing both before and after a particular moment. Sometimes, "I even think to myself, 'Oh, I hope I get this line across right.'" Though she has never completely forgotten a line, she has occasionally changed one a bit, which she finds unpleasant and nerve-racking—although she feels it has never visibly affected her performance, even momentarily.

Foreman sits behind a table positioned directly in front of the first row of spectators, running the sound and lighting equipment that provide cues for each performance. Manheim, for one, finds that having the director right there is very helpful. Because Foreman participates in the performances by occasionally yelling "cue" to indicate a change in activity, if someone forgets to come onstage, which rarely happens, he simply calls out that person's name and the performance continues without a break. His presence probably lends a certain psychological support—the degree, of course, would depend upon individual performers.

Foreman's performers have ample time to study the spectators during a performance since they spend most of their onstage time gazing directly toward the audience seating section, which is always well illuminated. Manheim states, "I look at the audience, noticing everything in great detail." During many of the earlier productions she was frightened to look at the audience and instead concentrated "on just myself being there, doing it blankly and not looking at them." Now, however, "I watch the audience very carefully and lots of times I try to figure out what people are thinking, especially people I know." When there is no one out there that she knows, she sometimes wonders, "Who am I doing this for?" and then, "I will single out someone who looks interesting and fantasize that I am doing it for that person." The presence of spectators is important for Manheim, and she enjoys it when people attend rehearsals, "otherwise it can be pretty boring." She feels that "it adds a whole dimension to performing when you can see the audience that clearly." This added dimension relates only to her own personal attitude toward, or enjoyment of, performing since, "I never change anything when an audience is bored, but it's pleasant when they're all 'with it.'"

The mood changed from scene to scene in *Rhoda in Potatoland,* but Manheim did not attempt to change her mood according to the scene. "Most of the time, I just do the things once I've gotten used to them." She feels that there is always a dissociation between the mood of a scene and the workings of her own mind.

> RHODA: *How far do you think my mind is now from the circumference of my head?*
>
> VOICE *(on tape): Compare. Her mind and your own mind. She is an actress in the play. But at this moment, her real mind is working just like your own real mind.*
>
> Rhoda in Potatoland

Indeed, Manheim's mind is in different places depending on the night.

> I remember one night these French people came, and I had made a translation for them to read—each line I would say, I'd look at them and I'd think

the line in French before I said it. It was most maddening actually. I was thinking in French all night long—I was completely double that night.

If she finds that she is not concentrating at a point when it is necessary to focus her attention, she "tightens-up" physically: "I usually contract my whole body, tensing up all the muscles." She sometimes refers to this procedure as "hardening" the muscles, and while her movement is not rigid, she rarely appears to be relaxed onstage.

When asked how she functioned in the earlier productions when the pacing was quite slow, giving the performance a contemplative quality, she stated:

> I can't stand contemplative states and I never get into them. Even though the plays were slower, I was never slower—inside I was working ten miles a minute. Sometimes, when it was very slow, I would think of each slow thing as a succession of very fast things—a very fast breaking up.

This state of consciousness relates to what many spectators have referred to as Manheim's unusually high energy level. It can also be associated with the nervousness she tries to consciously cultivate before each performance and her desire to concentrate on her "aliveness." Rather than focusing her attention on her performance as a whole, she is continuously aware of, and works on, every component part that makes up both the thought processes and the simplest activities, such as walking across the stage or looking out at the audience. She sees her performance as a mosaic of herself and she focuses on each separate part rather than, or certainly before, the whole. "I'm conscious of dissecting myself and showing the different parts—each part lives in itself but is also connected to the whole."

After each performance she does not feel she has to snap out of one reality in order to function in another because her state of being onstage is related to that of everyday life, only much more concentrated and intense. However, immediately after a show, she does notice fatigue, both mental and physical, that she does not notice during it. In addition, she frequently suffers from pain in her back that she is aware of during rehearsals but not during performances. This, too, indicates a state different from that of everyday experience.

Backstage, during a performance, she never talks to anyone, maintaining the same attitude and intensity offstage as on. In general, she does not relate to, or socialize with, the other performers in any production. She feels this is true partly because of a basic difference in attitude.

Even though they are all there by choice and they like the work and they are all doing their best, I think I'm more serious or concentrated because, for me, it's a matter of "life or death." In my life it's the most important thing and it's not for them.

In comparing performing to a life-or-death situation, Manheim describes one final state of mind she sometimes experiences onstage. It is a transcendent state but not in terms of meditation, because she does not consider it necessarily contemplative. "I sometimes think of myself as God onstage and in certain sexual experiences—it's like infinity. It's almost like dying in that it's bringing oneself to a limit where one gets a glimpse of the other side." She experiences this state in moments of extreme physical activity where she feels momentarily "outside of herself," such as the many physically rigorous fighting scenes in *Pain(T)* (1974). "This feeling of the other side I get mostly through a certain use of my body—I get it through pain and through extreme, physically difficult, things to do, or making them difficult." It is in this context that one can understand her tendency toward pushing the body into basically uncomfortable positions in her physical work. The particular qualities apparent in Manheim's total performance—her presence, energy level, and "the Rhoda look"—are the result of her work on specific techniques combined with the various "states of mind" she experiences during each performance.

What disturbs Manheim the most about performing is the fact that critics, when reviewing Foreman's plays, never discuss the acting.

> I was distressed at certain points this year when critics didn't mention the acting—it's as if there were no acting. I thought that was very unfair.

A few weeks after she made this statement, she was presented with an Obie award for her outstanding performance in *Rhoda in Potatoland,* at the twenty-first annual *Village Voice* ceremonies honoring achievement in Off Broadway and Off-Off Broadway theater.

Florence Falk

Setting as Consciousness

Richard Foreman—since 1968 playwright, designer, director, and producer of the Ontological-Hysteric Theater—teases, provokes, and deliberately frustrates his audiences with the overall objective of transforming habitual ways of seeing. Foreman's theater is designed to reorient the spectator's perception toward nonlinear forms of consciousness and toward a capacity for aperspectivity. Convinced that the world we call reality is only one among several possible worlds, and that its subjective correlate, our personal consciousness, is only one among several possible modes of experiencing, Foreman advocates that the way we see and what we see can and should be changed.

Foreman's settings are total, tightly controlled environments—architectonic arrangements of visual, auditory, and kinetic elements, both fixed and variable, static and dynamic, that exist within a given theater space and whose changing patterns are perceptible to an audience moment to moment. In each theater work setting conforms, metaphorically speaking, to the contours of consciousness and can be described as a sequence of notations by the artist as he (or "it") creates a work *about* consciousness. Every constituent assemblage in the larger, total presentation is meant to reveal a network of thought, feeling, and mood, mirroring each psychic occasion as it arises, surfaces briefly, and passes away. Each assemblage, in other words, is a pattern of consciousness that dissolves in the same way that fragments of thoughts slip through the mind.

Physically, Foreman's setting is the total loft-theater environment; the relationship between spectator and playing area is spatially conditioned. In this kind of "creative construction," to employ Bauhaus ter-

Originally published in *Performing Arts Journal* 1 (spring 1976): 51–61.

minology, setting is itself an art form in which every slight variation is orchestrated and scored.

Pandering to the Masses: A Misrepresentation (1975), the first play to be performed in Foreman's own loft-theater, will be used here to illustrate a number of design features characteristic of his work:

Separately defined spectator and playing areas incorporated as architectural features of the loft. Total space of the loft theater measures approximately 30′ × 150′. At one corner is the spectator area—seven tiers of wooden bleacher seats. To the right of the spectator area runs one of the supporting walls of the loft. Directly in front is the playing area, "framed" on the right by the loft wall (which also "frames" the spectator area) and on the left, by a long brown curtain through which performers make entrances and carry on flats and props.

Traditional proscenium or picture-frame arrangement. The audience looks directly and frontally into the setting as if looking at the surface of a picture. Though the picture is "framed," the actual performance place shows through in a deliberate Brechtian exposure of the anatomy of the stage. The construction of the setting is part of the real theater.

Variable and divisible playing area within fixed parameters. In *Pandering* playing space is narrow and deep (20′ × 75′) and divided into four general areas: Area 1, a flat surface (15′ × 20′) beginning at floor level directly in front of the first row of spectator seats; Area 2, a steep ramp (20′ × 20′) rising to a height of six feet from floor level; Area 3 (15′ × 20′), an elevated surface six feet from the ceiling; Area 4, a space from which performers enter and disappear, extending to the rear wall of the loft; and an additional area, stage right, tangential to Areas 1 and 2, with a large inset wall fan that occasionally revolves and draws the eyes of the audience into its orbit.

Concealed, or partially concealed, off-stage playing areas whose space is redeemed or made relevant by various devices. These devices include sounds projected from a distance, lighting fixtures that suddenly glow in dim recesses, and half-curtains that both reveal and hide concealed space, for example, the wall fan mentioned above.

Painted panels and screens used, in Kabuki fashion, to compose and recompose the setting and to emphasize shifting background-foreground relationships. Eisenstein noted in "The Unexpected" that by varying the dimensions of painted gates, doors, or houses on Kabuki screens one could achieve the cinematic effect of long shots and close-ups. Similarly, Foreman adds and deletes screens and panels from the set, continually altering spatial relationships. Before *Pandering* begins, the audience cannot see beyond Area 2; two painted flats, extending the width of the playing area, block further view. On each flat is a painted portico with two colonnades, giving the illusion that there is space beyond the porticos. Immediately thereafter the flats are removed and the length of the playing area is revealed.

Totality of composition. Almost everything in *Pandering*—walls, ceilings, floor, and many of the flats and props—is painted a deep chalky brown. Some fake flowers, set into various niches along the ceiling, seem to disappear into the brown paint.

Reproduction of props and set pieces in several facsimiles. Pandering includes several versions each of snakes, hobby horses, roads, and books.

Characteristic features and details. These include spectator bleacher seats, loft framing walls, curtains, columns, and Foreman's private area within the spectator area, where he sits during each performance operating tape machine, lights, slide projector, and giving cues—a virtuoso performance of the total theater person, since Foreman is simultaneously manifest as playwright, director, spectator, and performer. Several other features are notable: eight rows of strings stretched in parallel lines across the front section of the playing area, some white, some striped black and white; the lighting arrangement, which consists of three lit chandeliers, two white spots suspended over Area 2, and two additional spots focused directly on the spectator area, which is brightly lit before the performance begins, and during certain sequences of the performance; an oval mirror and two rectangular mirrors hanging on the right wall, tilted to "see" the spectator area and to reflect (and distort) different sections of the playing area.

In his essay "Meditations on a Hobby Horse," E. H. Gombrich writes that the common denominator between a symbol and the thing it symbolizes is not external form but function. According to this interpretation, setting is an adequate metaphorical substitute for consciousness only if its formal structure represents (i.e., *mis*represents) in some deeper biological sense certain relevant aspects of consciousness. Setting must, in other words, tell us something about the *function* of consciousness.

In works of art, one of the powers of metaphor is to present essences —that is, to embody artistically complex systems of meaning in relatively simple designs (e.g., serpent / evil, skull / mortality, Dove / Holy Ghost, etc.). But how does one embody something as elusive as consciousness? Clearly, one way is to compartmentalize it: in 18th century imagery, to make it into a house with many rooms, or in 20th century terms, to accept relevant categories like id, ego, super-ego, or preconscious, subconscious, or unconscious. Such categories may be said to "*mis*represent" consciousness in a meaningful way.

Foreman "misrepresents" consciousness by treating the setting as a totality that can be broken down into smaller and smaller survivable units. Before rehearsals begin Foreman designs the setting and has it built according to exact specifications. He then reduces each constituent to more elemental configurations, and holds them in place, as if under a microscope.

In *Pandering,* the familiar omnipresent taped "Voice" (Foreman's own) advises the fumbling Max:

> Try looking through the wrong end of the telescope.
> Everything looks sharper, doesn't it?

Max instantly obeys; presumably his seeing improves because smaller patterns, or units of information, are, for the first time, accessible to him. When we, like Max, look through the wrong (i.e., "right") end of the telescope to see the setting more closely, two principal structural features emerge more clearly: *density* and *framing.*

Density

The Russian Formalists considered "density" *(faktura)* as one of the most important qualities in their world of deliberately fabricated objects. The appropriately (i.e., creatively) *deformed* work of art was supposed to reawaken the perceptual power of the audience and remind them of the "density" present in the world around them. To help the audience, Victor Shklovskij, one of the leading figures in this literary movement, proposed the strategy of "making strange": the poetic image was always to be put into a new and unexpected context where it might be seen as if for the first time. In a sense, Foreman tries to make every image in his theater "strange," beginning with its fundamental shape—the setting—whose deliberately homemade appearance suggests to the audience that it has entered the artist's consciousness by the back door.

Density requires, first of all, the leveling of all theater elements. In a Foreman theater work, each given sequence is the construction of a kind of art-object in which all the elements—performers, machines, words, space, time, noise, things—serve as variables, and stress is placed on the formal relationship of these variables. As a result, theater experience no longer depends upon the intellectual super-structures that we normally rely upon to form our ideas and experience of the world and to match our experience with our perceptions of reality.

Visual patterns overlap, are superimposed, or repeated, so that the eye (and mind) wander over a constantly shifting theater-scape or field. As soon as one formal arrangement has, in Foreman's words, "seduced the senses," it is disassembled, rearranged, or discarded altogether (e.g., the swift changes of tableaux in *Pain(T)* (1974) and *Pandering*, particularly). The eye is not allowed to rest long enough on any form to be satisfied but is forced to readjust as forms continually shift and change location, and group themselves into new patterns that do not possess any stable center. As a result, foreground and background become ambiguous locations and perceptual experience tends to become more diffuse.

Performers. Performers are *merely* persons of masculine or feminine gender whose personal features are deliberately effaced. They may be said to represent certain states of mind or currents of thought circulating through consciousness; in other words, they are internalized representations of a consciousness engaged in an objective clarification of the self, similar in this respect to the six voices of consciousness in Virginia Woolf's novel *The Waves.*

Movement. In building up a setting, Foreman gives at least as much attention to its visual and kinetic dimensions as to the words in the text. Movements are either attenuated, e.g., Max and Celia's slow walk across the stage with walking sticks in *Particle Theory* (1972), or accelerated, e.g., numerous sequences in *Pain(T)* and *Pandering.*

Words. All words have equal value. We no longer grasp the center of a sentence, or feel the weight of a particular thought, accustomed as we are to the inclining gravity of a phrase or the growth of a sentence towards something. When, for instance, Sophia in *Sophia = (Wisdom): Part 3: The Cliffs* (1973) says, "I. am. your. equal. now," she is (flatly) stressing her right to think vertically as well as horizontally. She has equilibrated her thoughts, forcing even punctuation to comply, and thereby eliminating causal connections.

Music and sounds. Sound is part of the texture of each sequence; some sounds are pleasant, some are merely monotonous, and some are disturbing interferences that cannot fail to aggravate the nervous system. Foreman uses the buzzer, for example, like the prodding stick or electric fence that shocks dumb animals to wary attention.

Lighting. Foreman uses white lights to emphasize the formal relationship of planes and to expose the patterns that continually reconstitute his settings; thus, details are blotted out, shadows effaced, and the emotional impact of chiaroscuro eliminated. Implicit in the lighting schema is the redefinition of relationships and the conscious leveling of *persons* and *objects*. (The deliberate metamorphoses of person to person-object, and, conversely, object to some form of living matter, is frequent in Foreman's theater works.)

In Foreman's kind of reductionism, the terminology of the "old" theater also yields to the "new": actions become activities, indicating processes of shorter duration and a form that is more open-ended; and scenes become sequences, a continuous series of tableaux in which an activity, or series of activities, occurs. Sequence is also less definitive than scene which tends to convey a more coherent and inclusive relationship of a particular arrangement of formal elements and activities, even if the scenes as a totality do not form a plot. The "tableau" characterizes the pictorial arrangement of elements, animate or inanimate, or both, in a given area of space.

In sum, setting is a large facade, or series of tableaux. A tableau may or may not be static and may or may not require speech or a specific place or situation. Some tableaux have no movement (e.g., Max staring at the audience at the beginning of *Particle Theory*), some have small amounts (e.g., Rhoda, Eleanor, and Sophia on rocking chairs in *Pandering*), and some are dynamic, that is, they are organized around the element of movement (e.g., Leo pedaling the tricycle in *Pandering*).

Density implies simultaneity of perspective. Foreman has learned from the Cubists how to concentrate various perspectives into a single image so that space begins to feel real—almost tangible. In fact, Foreman's moment to moment or sequence to sequence theater form is analogous to the Cubist obsession with painting distances and spaces in order to make them as concrete as the objects or, in effect, to make the objects *surround* space.

One of Foreman's achievements is to create sensed space, space *retrouvée.* The spectator becomes conscious of space as an entity with form and density. By forcing the spectator's line of vision to shift continuously, the sense of space is altered. In *Pain(T),* for example, merely by changing the alignment of the wooden gate, a moveable part of the setting used in most of the tableaux, Foreman can vary the intersecting planes and thus lengthen or shorten space. If the gate is set on a line parallel to the audience in the forefront of the playing area, interior space seems to disappear, since the eye is stopped by the line and does not travel beyond the playing area closest to the audience. By setting the gate on a diagonal (left front to right rear), the eye naturally follows the plane to the obscure darkness of the righthand off-stage space. That blank area—simply space—suddenly comes alive because the eye has been directed to penetrate its darkness. Instead of watching only stage space, the spectator has discovered a nest of space that might yield some new discovery. Then from the rear of this unexplored darkness the spectator hears the taped voice of a tenor singing an aria. This is the reward for having reclaimed off-stage space.

When we look at an object, we unconsciously repress the ever-changing distortion of its form and texture in order to see its constant features. What Foreman does is arrest these shifting localizations so that the audience can see each object from several perspectives simultaneously. He is, in this respect at least, a Cubist playwright who, by constantly metamorphosing his formal stage images, constructs a theatrical whole out of a composite of shifting, intersecting planes.

Another way to corporealize space is to position and confine performers within a specified area. Foreman divides the stage into multiple playing areas with separate activities frequently occurring simultaneously in more than one area. Subdivisions across the width of the stage further increase the potential number of activities that can be carried on concurrently. Almost every activity is ritualized and reduced to a formal pattern. At given intervals of time these activities are rhythmically repeated so that there is an increasing sense of circularity; a kind of dream world is depicted. There is an invocation of man moving and acting because of his essence, pre-rationally, before the crystallization of thought, or before individuation has occurred.

Density allows for reproduction and convertibility of images. Just as boundaries between foreground and background shift and dissolve continuously, so do other elements. Another strategy to dim the sense of an "original" or primary gestalt image is to manufacture props and set pieces in several versions; consequently, each stage object has at least one (and usually more) mirror image that distorts the other, making it impossible to tell the "genuine" object from its "counterfeit" image.

In *Pandering*, the tricycle Leo pedals "dissolves" into Ben and his brother, the "Magician," astride the bust of a wooden hobby horse, which "dissolves" into the "high (hobby) horse" that Max mounts by climbing up a ladder. A triangular cloth "road" superimposed on another road (the ramp, Area 2) leads to the front door of a small-scaled model villa. Another "road" is constructed of wooden boards which are nailed together and attached to each end of a hobby horse. And all roads are "mirrored" by the string and wire "roads" strung across walls and columns in spectator and playing areas. Nor does Foreman stop here, for the dark stripes painted on the flats that represent Rhoda's room are themselves mirrored in smaller stripes painted on some of the wires and strings and other props, all of which may be said to "represent" other roads as well as planes of perspective.

Foreman also provides sophisticated reproductions of images from the history of art. To give just a few examples, many of the melodramatic tableaux in his musical *Hotel for Criminals* (1975) bear a striking resemblance to drawings by parodist Edward Gorey; in *Pandering*, nudes walking down a road (Sophia, Eleanor, and Rhoda on the ramp, Area 2) recall works by the surrealist painter Paul Delvaux; and the pie-throwing activity that takes place around a white oblong table could easily be a burlesque of many "last supper" scenes.

Foreman's settings indicate an awareness of kinetic art and use some of its features to suggest through movement the dynamic nature and interrelationship of all matter. Stripes and strings exhibit Foreman's concern with the quality of space and the nature of movement-through-space. His settings are similar to constructs of kinetic artists that include actual movement or else arouse the sensation of movement by means of optical and spatial illusion. At times, mass appears to dematerialize and the elements of his theater canvas no longer seem separately distinguishable.

Certain other correspondences with kinetic art appear in the texture of his settings: repetition of design or geometric configurations on painted surfaces, often in black and white or brown and black (the setting *Hotel for Criminals* is actually a study in black and white); geometrical subdivisions of space, designated in part by rods and wires; and superimposition of similarly patterned surfaces (e.g., prop against canvas flat, or wire and rod against flat, or moving forms against stationary background or the reverse, or, more generally, of the acoustical level against the visual level). What these techniques imply is the continuing sub-division of matter into smaller and smaller particles and their perpetual activity and interpenetration. Thus, the strips in Rhoda's room, which the Voice in *Pandering* assures us "have something to do with her body," may be thought of as frozen movement, or alternatively, as force lines or current.

Density is implosive. Air rushes into space, movement is toward the interior. The empty playing space in *Pandering* is, consequently, just that—an empty field. It is space anterior to an event (or a thought), for before consciousness develops there must be something to fill it.

As soon as *Pandering* begins, the panels confining space to Area 1 are removed, giving way to interior rooms, or rather, indeterminate locations. The passage from one space to the next indicates levels through which one must pass toward other forms of experience and knowledge. The composer, Stockhausen, tries to penetrate the deepest layers of sound, moving past the gestalt of an individual sound into its further subdivision, discovering there another nucleus that carries with it a new set of components. With each movement, the individual gestalt yields to a new multiplicity that yields to a new gestalt, etc. Similarly, the removal or addition of panels in Foreman's settings is meant to suggest new (and deeper) versions of consciousness. In *Pandering* there are countless references to secrets that must be revealed. To discover "knowledge," gloves (or bandages) must be stripped from hands, and clothes from bodies. In seemingly accidental or insignificant details the creative processes of art (and consciousness) unfold, removed from conscious detection. They are equivalent to secret messages encoded in secret letters hidden away in secret drawers everywhere: "Text, text, that's what counts," says Ben. "Not music, not ideas, not decor. . . ." He means that instead of listening to melodies, one must hear the vibratos, or in a painting, notice bold brush strokes that cancel out bolder outlines, or—in a Foreman work—consider even the stripes to see if they may lead to new planes of discovery.

Framing

Density allows us to see the force lines moving into and out of the "picture"; *framing* gives unifying vertical and horizontal support. Foreman uses the frame to isolate and hold an image in place, then shifts perspective (slowly or rapidly) to a new image. **The strategy of framing gives Foreman a continuous opportunity to**

maneuver the setting and thus outmaneuver the audience. In the field of the play, the relationship between spectator and performer is fluid, that between spectator and playwright-director, relatively fixed. Generally, the audience observes the presentation at an aesthetic and (hopefully) contemplative distance; occasionally, the audience *becomes* the play. In *Vertical Mobility* (1974), harsh white lights are turned against the audience during the "tennis game" to suggest the rather discomforting idea that the real competition is between performers and spectators. With their eyes concealed behind dark glasses, performers raise their rackets in menacing gestures against their opponents on the other side of the net (or frame). The inversion of spectator and playing areas, and of spectators and performers—spectators observe performers who are spectators, performers observe spectators who are performers—occurs frequently. It provides an example of infinite regress; each side of the frame becomes a mirror in which the other side can behold itself. An amusing burlesque of role reversal occurs in *Pandering* when three pairs of naked breasts, visible in turn through the frame of a small peep hole, invite the audience as voyeur to make aesthetic comparisons. The breasts are "eyes" ogling the spectators, scorning their libidinal curiosity.

Framing allows Foreman the possibility of modulating or combining various perspectives into a single focus. In one sense, the frame of the playing area punningly establishes a frame of reference, since it paradoxically offers a picture of reality outside itself. In another sense, the frame contains *all* of reality, since it is impossible to conceive of any segment within the framed area that does not *represent* or (to borrow from the *Pandering* title) *misrepresent* something, and which is not significant. Within the frame of a single sentence, the Voice in *Pandering* easily reclaims space that momentarily appeared to be dead:

> **Voice.** Dead space filled only by the return of Max who left but remained all the while thinking about the experiences Rhoda was having and wondering what he could say to prove that everything was happening the way it was supposed to be happening.

In the context of setting as a metaphor for consciousness, just as there is no such thing as "dead space" (Max leaves but remains all the while), there is no such thing as dead consciousness. "Dead space" is misperceived reality; space only appears dead because one does not see properly; or in surrealistic terms, it appears dead because one does not see into, through, or beyond it.

One repeated effect of Foreman's use of the traditional picture-frame stage is that of a succession of frames, one inside the other, in the fashion of Chinese boxes. In *Pain(T)* the Chinese-box device shows us how controlled an exposition of perspective Foreman provides the spectator. Halfway through the play two crew-members carry a large canvas on stage. It is the finished painting of an empty room. The painting is in tones of grey and brown, with receding diagonal planes that separate ceiling from walls and floor, moving from left front to right rear. A door in the left forefront wall is similar to the door in the setting that Max tries to open. At each side of the painted canvas are two white columns, like the two poles that frame the playing area. The resemblance between canvas and set is unmistakable.

Shortly, a second, smaller canvas, which duplicates the first, is brought on stage so that, simultaneously, there are three versions of a room—three separate realities. But even without close duplication, the point is well made. Outside the setting of the play is an even larger canvas room that includes the spectator—an expanding process that could be extended *ad infinitum.*

Miniaturization is a central feature of framing. As with Max's telescope, the image is adjustable and can be contracted or expanded. On one side Foreman opens the frame so widely that it receives a holistic image; on the other side he contracts or miniaturizes it to concentrate the image. In *Pain(T),* performers carrying miniatures in their hands cross the stage area on a plane parallel to the spectators who must squint to see such small-scale images. There are also larger versions of these miniatures placed on either side of the set and left there during the course of several sequences, or else carried on stage for a specific tableau.

Pain(T) is, in fact, structured, both literally and figuratively, so that the entire play resembles a miniature. It miniaturizes or compresses into the space of the playing area the history of perspective in Western art. The play moves forward into the present in a series of tableaux that feature the role of the nude in the history of painting—from Titian's naked women to the present.

Framing exists even in wordplay. Foreman's primary pun is the "pain" in *Pain(T)*. The double reference (pain, paint) is emphasized throughout the work. We can, if we choose, see only the pain in pain(t), or, with the assistance of frames of reference and of perspective, we can see both. Similarly, in *Pandering to the Masses: A Misrepresentation,* we can see "misrepresent" but possibly *miss* "represent."

Framing reveals the condition of aperspectivity. Framing is a device to accommodate multiple perspectives. There is no indication that Foreman prefers one perspective to another.

It is best—and this appears to be Foreman's most important lesson for the spectator—to see both possibilities *within* the larger, aperspective frame. The perspective vision is, in its most encompassing metaphysical sense, imbalanced, disassociative, fragmented. It is always a partial condition, in which the part prevails over the whole. It is also person-centered. The danger of perspective when we focus on the "pain" in pain(t) is that the viewer regards the space between himself and the object as something to be overcome. To reach the object, space must be mastered. But the subject is then trapped by the object.

The mirror, used by Foreman in *Pain(T)* and *Pandering* especially, consistently thwarts perspective. In *Pain(T)* the kaleidoscopic use of mirrors distorts and fragments reality. Placed on the set, or held by performers, mirrors play havoc with lines of perspective and refract and multiply the worlds available to us. The mirror, a symbol for consciousness in Eastern religious thought, equilibrates input and reflects reality without deletion or embellishment. But if the mirror is trained to reflect different and unaccustomed areas, then it can defy the limits of perspective reality with ruthless objectivity.

In relation to consciousness, the notion of aperspectivity might be expressed as left and right cerebral hemispheric integration, or as a mature and balanced accommodation of linear and nonlinear consciousness. Aperspectivity, encoded genetically into the structure of consciousness (and exemplified in the Chinese-box structure of Foreman's settings) does not negate more traditional forms of consciousness; rather, it both integrates and supersedes them, with apparent loss, however, of their separate autonomies.

Indeed, aperspectivity cannot be expressed except by reference to perspectivity. The contours of newness are concealed or misrepresented unless its forms are set in relief against familiar, older forms. Foreman takes delight in this abrasive and ironic interplay of forms: he continually establishes the dialectic between old and new, using one frame to cancel and overcome one point of view, be interpreted as merely book knowledge, or linear thought; from another point of view, however, the *frame* of the book opens into other forms of experience. Even the "old" theater form is embedded, as parody, to be sure, in the structure of the "new." In *Pandering* the dumb show (the play "Fear"), a relic of the old theater, is structured into the center of a play, while throughout most of the performance, the Voice persistently reminds us that we are still watching the "Prologue."

Whatever excitement exists in Foreman's theater arises from the tension between these polarities that are established—horizontal and vertical and inner and outer force lines, old versus new structures. Yet to be discovered, however, is whether Foreman will create a more convincing dialectic between the subjective and objective sides of his own consciousness. So far, we have only witnessed an extreme objectification of consciousness (Foreman as "witness" to himself in the person of Max; Foreman, the taped Voice, matter and anti-matter at once). But the frame can conceal as well as expose. One wonders whether Foreman will eventually use the metaphor of setting as consciousness to offer up not only density but profundity as well.

Guy Scarpetta

Richard Foreman's Scenography

General Scenography

It is impossible to isolate one purely scenographic element in Foreman's work. His scenography, in a strict sense—his scenic "writing" or design, concerning the space, decor, and costumes—never ceases to overwhelm its traditional functions. There is, without a doubt, one major singularity to Foreman's style: scenic "writing" is everywhere. Spatial tracing is not simply a phase in the spectacle's preparation—it persistently shows itself off and sets the resulting production in motion.

Foreman's scenography is not a code that one can contrast with others in polyphonic theater—it cuts across all theatrical codes. Foreman's art, at base, is an art of "contamination." The decor intervenes in the action; it is a "performer." Characters assume a purely spatial, rhythmic, decorative function. Costumes and props play a dramatic role. In this manner, Foreman explodes the classical oppositions on which theater has been based, oppositions between decor and action, between animate and inanimate, between the accessory props and the essential ones, between scenographic space and spectacular time. His theater requires new instruments of analysis; it becomes necessary to think in terms of energy, tension, lines of force, and variations of intensity.

After the "Avant-Garde"

Foreman's first plays participated in the prescribed goal of the "avant-garde" (the principles of which were paralleled in the cinema, painting, and literature of the same period). His work particularly reflected the aims of those who wanted the "hidden work" to be shown,

Originally published in *TDR* 28 (summer 1984): 23–31. Trans. by Jill Dolan.

who felt the production process should be present in the product, and who felt the resulting production should attract attention to itself as the deconstruction of its process. Foreman's work rendered the scenography visible; it created transformations from the perspective of the scenic space. The presence of strings drew a large axis of the movement and situated the actors as elements in an abstract canvas, systematically utilizing the props as components of an unearthing process (they functioned as "crutches" or "hindrances").

More recently (since his turning-point play, *Blvd de Paris*), the "Brechtian," avant-gardist principle has become considerably more flexible. Although Foreman continues to question the materials of traditional scenography, usually held secret, the modality of this materialization is less mechanistic. Thus, in the three pieces presented in France (*Café Amèrique* in 1981, *Faust ou la Fête Electrique* in 1982, *La Robe de Chambre de Georges Bataille* in 1983), the strings persist, but without their omnipresent, permanent character. They are no more than discreet traces of abstract scenography that outline the plays' internal movements instead of underlining them. The modifications of the space (in which the actors continue to participate) are more exact. They have ceased to invade and subordinate the action. The props, in a certain way, have assumed more autonomy through their rapport with the scenography. They have a properly plastic existence and intervene less gratuitously in the dramatic action.

For all this, Foreman has no intention of returning to illusionistic art, an art that would efface his process (as is the case with Robert Wilson, for example). This exhibition of hidden scenography is the principal effect of the representation—it participates with a suitably dramatic effect, without substituting itself for the drama. Phrased another way, it is now impossible to suppress the perceptual mutations that the "avantgarde" introduced, but its artists no longer seem intent on being locked in their process.

Space in Transformation

Scenographic space, for Foreman, is never a given beforehand. It doesn't constitute a set frame on the interior of which the action would be able to unfold; it modifies itself unceasingly—particularly in its dimensions—in proportion to the evolution of the action. In some limited cases, the modification even constitutes the action. The scenography, in sum, is mobile. The set is likely to become hollow or shrink, to swell or contract, to divide or multiply.

The theater loft in which Foreman directed four of his New York pieces was, from this point of view, very significant. It permitted a deep set that became a device capable of diminishing or deepening at will. The choice of this space indicated, in some way, that Foreman's theatrical art was post-cinematographic art, or at least an art likely to respond to the cinematographic challenge. Foreman was able to include "depth of field" with these methods, and eluded the flat frontality inherent in traditional scenography.

The more classical stage of the Théâtre Gennevilliers, which welcomed Foreman's realizations in France, adapted less easily to depth of field. Without being an "Italianate" stage, it nonetheless imposed the constraint of a large space with reduced depth. Foreman maintained the effect of permanent spatial transformation within this frame in three ways: (1) by the same use of actors, conveying in their displacement the dispersion or splintering of the playing space, which produced an impression of spatial expansion; (2) by the presence of vertical retractable gates or barriers slightly resembling mobile cells, pieces one might find in an urban space to control crowds (as in *Faust ou la Fête Electrique*); (3) by the eruption of props of unusual dimensions (a bar, animal puppets, instruments of motion, etc.) that produced an impression of breaking open the stage. In addition, the manipulation of the space's height, or its verticality, was apt to supply the illusion of depth (by the lighting method, or the presence, in *Faust*, of simulated-stone blocks suspended in the air, likely to descend onto the stage at will).

The alteration of classic space was not, therefore, abandoned. If, essentially, the scenographic frontality was not radically denied, it was nonetheless given the impression of being one possibility among others, menaced, always at the point of being disturbed. Foreman's theatrical space—which he does not hesitate to refer to as psychic space—is above all an unstable space. It is mobile scenography of vacillating volume and vertigo.

"La Chauve-Souris," or Foreman's Reappropriation of a Traditional Space

In the mise-en-scène he created for the Paris Opera's production of Johann Strauss's *La Chauve-Souris (Die Fledermaus)*, Foreman was not the absolute master of the scenography. He did not direct the design of costumes or decor, nor did he see the playing space in advance. Nevertheless, the resulting space incontestably bore the mark of

Foreman's singular style, as if once again, his unwillingness to depart from it had served as a provocative challenge.

The production had three principal characteristics. The presence of vertical, movable bars delineated the spatial restrictions of the play, which had been edited—the last act in this production was divided, distributed between a prologue and an epilogue. Where the traditional mise-en-scène of the operetta effected an enlargement of the space (for example, in the Ball scene), Foreman preferred to enlarge the set vertically. He revealed frescoes inspired by Ingres' *Bain Turc (Turkish Bath),* a swarm of nude, entangled bodies and several suspended angels likely to descend toward the stage.

The effect of breaking open or disrupting the space was produced, finally, by the movement of the bodies. The ballet was drawn little by little toward an angry *bacchanale,* with the turbulence of animal puppets and roller-skaters in Tarzan costumes that unbalanced the classical arrangement of the waltz. The lights played their prescribed role in this effect: The searchlights directed at the spectators—scandalizing certain purists—provoked a violent sort of perception, destined to restore and accentuate the work's intense virulence, which had been flattened by the proprieties of tradition.

This scenographic treatment, by which Foreman reappropriated the space that was imposed on him, converged toward a deliberate effect. It reacted to the "minor" character of the operetta, not by treating it as a major work—as an opera—but, to the contrary, by making it even more minor. Foreman placed its delirious, feverish, frenetic dimensions in relief and drew it closer to circus.

Literalization

Brecht proposed a "literalization" of theatrical space by inserting readable texts, montage effects, and a collage of written elements into his work. For Brecht, it was necessary to provide a sort of critical counterpoint to the representation. Foreman retained this process, but changed the direction of its function: Several letters—those of the play's title—were hooked to the wires that dominate the playing area in *Café Amèrique.* In *La Robe de Chambre de Georges Bataille,* ciphers were disseminated through certain elements of the decor.

These signs are not direct signification; they indicate a "literal" space without being subordinate to it. Foreman's method is not didactic. These signs signify nothing but their abstract function. By discreet contamination, they designate as signs all the other elements in the set. The spectacle furtively evokes the page of a book, or an outline in a note-book, the trace of a sort of algebra-in-abeyance. More than being Brechtian, Foreman is closer to the utilization of signs—letters or ci-phers—by Cy Twombly, for example, in which a likeness with Fore-man's literalization may be found. The presence of these signs is not aimed at the spectators' comprehension, but at their *perception.* It sug-gests that the "delirium" it assists also possesses secret logic, cipheriza-tion, writing.

Without doubt, the same diversion of the Brechtian scenographic process can be seen in Foreman's use of photographs. Where Brecht's photography possessed a didacticism—to eliminate representation by the eruption of the "real"—the hundreds of photos stuck together end to end in the upper part of the decor of *La Robe de Chambre de Georges Bataille* aim especially for a perceptual effect. It feels as though the play is about the photographs, although their accumulation gives overall a purely abstract impression. There is a sensation of entanglement, of proliferation. The hundreds of cut-out bodies, stuck together and mul-tiplied, refers to a sort of swarming organic chaos, accentuating the ir-rational, fantastical dimension of the play on stage.

The Props

Foreman's theater, more than any other, consumes the objects that he brings together with an intense rapidity that often resembles a veritable rage of destruction. Props and costumes have no value other than themselves. They appear only to be immediately devoured, swal-lowed up in the spectacle's hectic rhythm.

Their enumeration reveals their insensible character: decapitated ba-bies in celluloid, forceps, beef bones, a dentist's chair, cooking instru-ments *(Café Amèrique);* bars, a sarcophagus, meteorites, toboggans, ser-pents *(Faust);* skulls, restaurant materials, candlesticks, animal puppets *(La Robe de Chambre de Georges Bataille).* These different objects have no meaning—the task of finding their symbolic value would be ex-hausting—but they have functions. They focus or impede the actor's action; they call forth incongruity (an echo of cinematic burlesque that also includes a great consummation of objects); and they create an im-pression of disproportion and permanent disequilibrium.

In sum, it is not the scenographic dimension of the props that is pri-mary—their "form," in other words—but the way in which they are placed in a whirlpool, a scansion, an uninterrupted consumption. It is as if the profound goal of Foreman's theater was to ruin representation, to destroy that which he brings together and uses. Space, and the objects that rise into view there, are literally consumed by the spectacle's infernal rhythm and tempo.

The Theater as Pictorial Art

Scenography is traditionally the place where dramaturgical art touches drawing and painting, a sort of minor pictorial art with restricted, purely utilitarian functions—a preliminary, preparatory art effacing itself in the representation it makes concrete. This is precisely what Foreman radically reverses. In his style, pictorial art is less a point of departure than a result. With Foreman, it is the total spectacle—not solely the preliminary scenography—that confronts strictly pictorial problems: the disposition and dispersion of usable signs, effects of equilibrium, echoes, distortion.

Nothing is more significant on this point than Foreman's use of actors. He leads them surely, carefully, to "play a role"—or at least to enter fragments of roles—without ever forgetting that their bodies are also plastic signs. Their entanglement, their rebounding, their way of forming processions and garlands, their placement in "tableaux," all correspond to abstracted, disordered configurations in which the bodies are merely "organic." Lines of force, tension, dissemination, convulsions, including the most concrete (the body) in the highly abstract (the axis, directions, masses), all are elements of Foreman's universe, and suggest strictly pictorial experiences. They recall Hieronymus Bosch (by the confusion, the art of incongruity), Duchamp (by the visible emergence of a "fourth dimension"), and Picasso (by the systematic twisting of perspective, the art of decomposition and montage).

This theatrical pictoriality, once again, gives rise not only to the scenography. It is the general effect of theatrical codes—voice, rhythms, displacements, resonances, contortions—and not only the visible, stable signs of scenography that participate in this incessant composition, always in the process of undoing and re-forming themselves.

To treat everything that concerns theater pictorially further estranges it from drawing or painting: Such is Foreman's challenge. The theater, then, becomes sceno-graphy: the graphics of the stage and the stage of graphics at the same time.

Mel Gussow

Celebrating the Fallen World

As a playwright and director with a philosophical bent, Richard Foreman is a practicing metaphysician in the experimental theater. During the last twenty-five years, his plays—forty of them since he originated his Ontological-Hysteric Theater—have repeatedly analyzed the imbalances between art and life. Even while he has branched out to become a director of opera and plays by others, his contribution remains instantly identifiable. Whatever he does, he leaves sight and sound tracks as his signature, and his art has always had a deeply intellectual foundation.

In rehearsal at St. Mark's Theater with his new play, *My Head Was a Sledgehammer,* Mr. Foreman is at the electronic controls, interjecting suggestions to his actors and orchestrating the mechanics of the enterprise. With a mournful mien that extends from his eyes to his mustache, he is a younger, avant-garde doppelgänger of Broadway's David Merrick. As the author fine-tunes the performance, a spiderlike chandelier revolves like a fan in a Singapore hotel; string—a Foreman trademark—crisscrosses and stratifies the stage, and Oriental carpets cushion the walls. Actors in black dunce caps rush by as a professor prepares to lecture at a blackboard.

The source of *My Head Was a Sledgehammer* is Friedrich Hölderlin's fragments for a play about Empedocles, the Greek philosopher "who was destroyed because he tried to bring down to people a kind of truth not meant for humans." Mr. Foreman's version is, he says, "a gloss" on the original, and deals with a professor who "through silliness and weirdness tries to introduce his students to the poetic method."

Coincidentally, Eric Bogosian, who is a friend of Mr. Foreman's, will

Originally published in *New York Times,* January 17, 1994, CII.

soon open his new one-man show, *Pounding Nails in the Floor with My Forehead.* Mr. Foreman jokes, "I thought we might have a Jack Benny–Fred Allen feud about who stole the title."

Serendipitous Adventures

In his unwavering career, Mr. Foreman has been converting "clouds of language and impulse" into alchemic theater, from his early plays in which his muselike heroine, Rhoda (played by Kate Manheim), undertook serendipitous adventures, to last season's *Samuel's Major Problems,* in which his authorial surrogate was hounded by questions of mortality. Beneath the surface somberness is a comedic disposition. The plays owe as much to vaudeville as they do to existentialism.

Bizarre humor remains endemic to his work, although these days Mr. Foreman is, for personal reasons, more dour. His father died last year and Ms. Manheim, who is the playwright's wife, is very ill and doctors have been unable to discover the cause of her malady. Some time ago, she stopped acting. At 56, he is feeling more contemplative.

Asked to comment on his plays, he apologizes for risking pretension, and says: "If you were going to ask Heidegger what his next book was going to be about, he would say, 'About Being.' Well, all of my plays are about that." He might have added that the subject is also about being Richard Foreman.

The plays, which are continuing chapters in the serial of his mind, start with half-page scenes in his voluminous notebooks. Periodically, he harvests his entries for material that is thematically related. During an intensive eight-week rehearsal period, he processes these excerpts through his directorial imagination, and searches for a scenic environment "in which those lines of dialogue can have some reverberation." He continues: "People might be amazed if they saw how carefully we worked, how much material is thrown out and how much we change. Usually my cast is in agony because they think I'm throwing out all the best parts."

In any traditional sense, his art lacks narrative. Things happen and then happen again. Scenes can be switched, without disturbing the work's intention. He is "much more interested in juggling ideas than in telling a story." To illustrate the approach, he recalls a statement made by John Gassner, who was his playwriting teacher at the Yale School of Drama: "Richard, you have talent, but you have one problem, which is that you get a strong dramatic effect and you just want to repeat it and repeat it and repeat it." Eventually, the playwright took that criticism as a compliment. That was *exactly* what he was trying to do, to find an effect worth repeating, "the one effect you never tire of."

Before going to Yale, Mr. Foreman graduated magna cum laude from Brown University. He has always had a scholarly streak, but only gradually was he able to accommodate it in his theater. At first, he wrote plays in a Murray Schisgal vein, one of which was considered for Broadway as a vehicle for Alec Guinness. The English actor declined the offer, saying that he liked the play but felt that he was the wrong actor to portray an overweight, middle-class Jewish man from the Bronx. In retrospect, Mr. Foreman says, "With his great chameleon quality, I suppose he could have played the part."

The author's career charged in the opposite direction. Taking inspiration from the independence of underground film makers of the 1960s, he asked himself what he wanted to see onstage. He had an image of a nonsequential, idiographic theater in which "people faced each other across a space and said a few abrupt things, then moved and said a few more abrupt things."

Beneath the apparent anarchy was a sense of order and mystery, even an Aristotelian logic. From the outset, his theater has had a strong literary base, drawing on works of philosophers, with whom he has one-way dialogues. Oddly, he has always had an ambivalence about his profession. Given a choice, he might prefer to be home writing or reading Eric Vogelin on the history of consciousness. "I've always been a rather withdrawn fellow who is occasionally dragged out into society to put on a play," he says. "I'd be a hermit if it wasn't for theater." Compensating for his shyness, he is an integral part of the performance, with his tape-recorded, sepulchral voice offering wry asides on the drama.

For Mr. Foreman, as a poet-philosopher, the plays are meant to be instructive. He explains: "Art is trying to redeem, to learn how to dance with the problematic aspects of the world. It's easy enough to imagine a beautiful world, and to celebrate it, but I would rather learn how to celebrate the fallen world we live in. People have looked at me awry when I say that I think my plays are pictures of paradise. There are obviously plenty of very aggressive, unpleasant elements in them, but I would like to think they are subjected to a kind of aesthetic message."

His distinctive brand of performance art communicates with a seismic, prop-dominated theatricality. From the beginning, he has tried "to build the potential for unexpected collisions into the physical materials on stage." The design—aural as well as visual—is fraught with peril, especially for the actors. "If a table has a fat leg or wobbles in a funny way, that automatically suggests trouble," and breakaway furniture results in breakaway comedy.

Despite his devotion to a theater of ideas, the playwright has an active interest in more popular aspects of culture. As an admirer of Jule Styne, he harbors a dream of directing a production of *Gentlemen Prefer Blondes,* his favorite musical. But he has no interest in reviving his own plays (although *Dr. Selavy's Magic Theater,* a musical collaboration with Stanley Silverman, was brought back some years ago). The works exist in the moment for the audience that is watching.

Reflecting on his theater, he says, "To me, art is really 99 percent courage, the courage to follow your vision, and to remember what your particular vision is. It's a struggle not to let your mastery take you down the easy road, and in spite of what some people think, I have a mastery in certain areas. I have to cast that off so that I'm back in a naked condition confronting the material of my life."

Question of Finances

Serving as his own producer, he uses income earned from his outside directing assignments as well as foundation support, and loses money on his shows. "I don't make a living from the theater, but that's fine," he says philosophically, adding, "There's something missing from my life, in spite of the fact that I've been very lucky in many ways and I've lived with the same lady who I love very much for almost thirty years now." His theater is his relief. "I only do it to feed myself," he says, "and I hope that someone else needs the same food."

Marc Robinson

A Theatre of One's Own:

The Mellowing of Richard Foreman

Against all odds, it's a persona we've grown to love: Richard Foreman, woolly philosopher of the avant-garde, detached and rigorous, with a gravelly voice that speaks only in paragraphs, committed to a life spent turning abstruse theory into theatrical gold. And the saddest looking man in show business.

The face is still saturnine and, on a recent visit to his Soho loft, one could see philosophy stuffed in the sofa and piled hodgepodge on the coffee table, including one notebook on which he's emphatically written "Theory." But as anyone who sees his latest play or just talks with him a while soon understands, the remote persona no longer quite fits. For starters, he's unabashedly contented.

"I used to think of myself, as any good avant-garde artist would, as this angry young Turk, making hard-to-relate-to work. But I'm in my early fifties now, and I just want to reflect the mellowness of this stage of life. I want to be in contact with ultimate things, to feel my way back into deep rivers of feeling. I'm amazed people still think I'm trying to shock them."

If anything shocks his audiences these days, it may be such unfashionable words as "ultimate things" and "feeling." Foreman knows he probably sounds corny to the self-serious, and the earnestness about spirituality may disappoint those loyal fans bewitched by his theater's baroque splendor, its obsessive materialism. Encountering a skeptic, he'll say he's *always* been a closet religious writer—but until lately wouldn't admit it out of shyness, and an understanding of how easily such private impulses are cheapened. And so, as he does with every equinox, Richard Foreman is threatening to retire.

Originally published in *Village Voice,* April 23, 1994, 92, 94.

He himself says he probably never will—if only because what he writes is meant for the voice and needs an arena. What he will do, however, is leave behind once and for all one form of theater-making. "I used to be excited about ordering hordes of people around the stage," he says, only faintly wistful. "But now, I've had it. I'm no longer interested in bleeding armies. I want to return to the hearth, with one or two loved ones, and make something quieter."

It's an urge that finds poignant expression in *Eddie Goes to Poetry City* (at La MaMa through April 28), where Foreman's hero flees a deadening society and searches for a home so secluded and comforting that he just may see his personal god. *Eddie* is one of Foreman's most emotional works, a deliberate choice, he says, made after preparing the play's Seattle production last winter. "Maybe it was the pastoral setting," Foreman says, "but that version was more like a Gertrude Stein landscape play—looser and very abstract." Now, in New York, *Eddie* seems to have moved indoors, into a more time-bound world, and its tone is that of chamber theater, a form he started experimenting with several years ago, with *The Cure*. Foreman has retained only a fraction of the Seattle text, and thoroughly reimagined his characters. Out West they were "decadent, childlike angels operating in a velvet-carpeted, adult, slightly pornographic playground." Here, he says, they're approaching a hushed, Edward Hopper–like demeanor that goes with the setting's airless office and dark, warm analyst's den.

Foreman's retreat into his own lair will have to wait, for he's off to Lille this month to do *Don Giovanni,* and he's also promised to direct William Finn's new musical next year. But in the meantime, he's looking for a charmed room like the one he gave to Eddie, a theater of one's own, no bigger than his parlor, where he can summon trusted actors—along with a sound man, of course (some things can't be sacrificed)—then let in no more than a score of spectators, and play. Those who fondly remember the days of the Ontological-Hysteric's intimate performance loft will be happy, no doubt—but Foreman is careful to note the difference. He's started writing long narratives in which the language takes precedence over everything else, even his elegant designs. Last year's *Lava,* with its essayistic form and mournful ruminations about the inadequacy of language, was the precursor and challenge to the work he's making now. Eddie's story is but one of many parables—among them *The Amateur Genius* and *The Mind King,* all yet to be staged—in which, for the first time, Foreman says, he feels himself entering totally into the act of writing.

"It's been very difficult for me to reach the point where I'm willing to admit that I write like taking dictation—that all the best things that come through me I don't control." For Foreman it's less a confession than an epiphany. The psychoanalysis in *Eddie* could be a reflection of Foreman's own quest for an ultimate relaxed state—a condition of mind when ideas and responses surface without the aid of even the most provocative theory. "If there's any sadness in my plays, it's the sadness of someone who sees how much time you spend covering over the pure music that's really you and that you're meant to sing."

Foreman still includes himself among the hopelessly distracted. "I'm basically uptight, conservative, and unadventurous, always busy with meaningless pursuits. And so my big problem is feeling free. I'm too repressed even to get up on the dance floor. The big thrill of my life was when I directed *Threepenny Opera* and had to demonstrate choreography to Raul Julia. I actually could do it!" Now even the language dances: in his newest writing Foreman sees a sensuousness that never seemed possible until now. "The tenses, subjects, and objects are continually shifted and sliding, as though the plays were machines in which there was too much oil in the gears.

"Of course, I haven't any idea how to stage this," he happily concedes, and no wonder—such writing, with its whirlpool language that doesn't always carry you forward, defies the theater's blasted injunction to keep only the action moving fast. And as the linguistic patterns become even more complex, the new work also challenges Foreman's hell-zapoppin' style of directing. "I wish that when you went to the theater and something interested you, you could just call out, 'Wait a minute! Let me go out for a drink and think about that for a while, and then I'll come back and you can continue.' Or even better," Foreman goes on, "I wish I could figure out a way to make a play that you could just have in your house, so you could relate to it as to a book."

Indeed, Eddie leaves theater for a city of poetry, and all of Foreman's recent work seems to yearn for a lyrical suaveness and grace. If asked what experiences surrounded the making of *Eddie,* he'll list book titles sooner than he'll recount autobiography. Reading *is* Foreman's autobiography—and during the years of work on *Eddie* he kept returning to three poets: Zanzotto from Italy, Sweden's Ekelund, and the Romanian Lucian Blaga—esoteric, perhaps, but not to someone who will go to the theater only five times a year (Foreman's maximum) but doesn't let a day go by without dropping into a bookstore. How pleasurable for him, then, to anticipate the publication of two new collections of his own: *Love & Science,* a book of librettos named after the musical can-celled by the financially ill Music-Theatre Group, and *Unbalancing Acts,* the five most recent plays, which Pantheon will issue this winter. For an artist who looks upon theater as an act of making "places where the world is as it should be" for the short duration of a run, the literary breakthrough is more than gratifying: he wins the peace of mind of knowing he can't be expelled from a published paradise.

Such a home is especially important to Foreman as he faces theatri-cal uncertainties. He's worried audience won't come to *Eddie*—attend-ance was disappointingly low for the last couple of pieces—and he's nervous about funding. "I've discovered I'm not a fundable artist," he says, "because I'm no longer 'emerging'! And I don't have a gimmick—I'm not planning to take Yankee Stadium and choreograph thousands of people, so no one's interested. There's no glamour to me."

Hoping, perhaps, to change that, he says, "Now don't you want any gossip?" How can you ask for dirt from a man who just finished telling you that "even as a little boy I worried about the fate of my soul"? Later, with a sly smile, he announces that he and Kate Manheim, his long-time companion and favorite actress, have finally married, just last De-cember. For the erstwhile young Turk, who till lately never voiced so emotionally the need for spiritual and artistic homes, such a gesture is positively scandalous.

II Reception

Arthur Sainer on *Total Recall*

It's not often that I urge readers to break down doors to get into this or that performance, for I don't consider it appropriate to use these pages either as consumer referral stations or as marketing bleeps on the theatrical scene. But Richard Foreman's *Total Recall* seems to me a work of such magnitude that I feel it's legitimate to urge you on to it during the brief time left to it at the Cinematheque.

I also should warn you that the work is austere, overly long (three hours without intermission), and at times impossibly clumsy. When it isn't working, its failure is due to a breakdown between concept and physicalization. There's a peculiar siphoning of energy from the flesh of the field which is partly Foreman's intention, but it tends to get away from him, or to get to him as it's drained from performance to creator. When the production is off, it's off because there's too much head.

But it's a cunning, forceful, and brilliant work. Foreman's focus is on what begins as the commonplace. A plain kitchen table, a couple of plain wooden chairs, a plain bed, a plain tree, a plain husband. Also a wife, a dog, a small girl, a lot of plain lamps.

Also a goddess. Also an Uncle Leo.

The characters are barely heard. That is, they're heard distinctly, in monotones, on tape, but the live characters, in even flatter tones, overlap the taped dialogue with a word or at best a phrase; the words are almost futile but calculated resonances.

The play is punctuated, and the audience assaulted intermittently, by what sounded to me like the workings of an electric saw brought up to the highest decibel.

Originally published as "A Hot Tip for a Bitter Winter," in *Village Voice*, January 21, 1971, 55.

67

The visual field is dominated by threshholds—doorways, windows, entrances to closets. The goddess is seen, over and over again, in the closet. The husband is seen on several occasions half into and half out of a room. He is also seen entering the closet, and leaving the closet. He is seen looking out the window, he is seen half-way out the window, he is seen wholly out the window. The kitchen table is also seen near the window, it is also seen outside the window. The tree is seen out the window, it is also not seen out the window in those moments when the stagehands have "disappeared" it. The sense of tableau is strong. What is seen has come into being because energetic forces have prepared for it and it is seen predominantly in a state of rest, it is located and benignly percolating in its state, in its being. We see tableaux, pictures, arrangements of pedestrian life confronted in its being.

Foreman also uses lamps—everywhere, hanging from the ceiling, almost but not quite touching the kitchen table, resting in the hand of the goddess, carried by the wife, attached to a dressing table, everywhere. If the threshholds are places which contain the possibility of allowing energy to pass through or to rest, the lamps seem to be that energy that not only isn't us but is higher in density than what we haven't been able to will ourselves to as yet be. The lamps are not only homely possibilities but languid affronts.

Foreman says, "Clarity is not having ideas but putting space enough to see what arises." And says, "If every moment can be clear and only itself, the stasis becomes a tremendous release from fantasy and storytelling. And delight is in everybody's head instead of in objects and various postulated adventures."

Total Recall is a fussy, marvelously located examination of pictures of states of being. P. Adams Sitney, in his long "preview review" which is handed out at the theater, believes the work could easily go on without us. "Like canyons and rivers his plays repeat their existence." In that sense, the examination is like a child's examination, "a child's play, it's for itself, it doesn't know us, it's busy repeating its kind of aliveness."

Michael Smith on *Evidence*

Evidence, Richard Foreman's latest Ontological-Hysteric Theater work, is playing through this weekend at the Theatre for the New City. This has been a rich couple of weeks for art theater, what with the series of events at the Whitney, Foreman, a new manifestation by Robert Wilson/Byrd Hoffman, Meredith Monk posing and singing and moving a little, and I don't know what else. All these people come to theater with some other art's sensibility and social life, making performances that don't resonate to the traditional theater vibes, that to those expecting drama or entertainment often seem (and may be) perverse and uncommunicative.

I saw Foreman's first piece about four years ago at the Cinematheque and then missed the intervening work, including *Lines of Vision* and *Hotel China. Evidence,* according to a repeated tape-recorded announcement, "consists of material that could not be made into a play." Specifically, it is two hours of stage animations from the notebooks in which Foreman's earlier works were written and sketched.

The theater set-up looks grim, what with bare walls, bare light bulbs, found furniture in a spare, rigid arrangement, taut white cords crossing through the air, a blackboard on the upstage wall, most everything painted in dull browns, the whole effect anti-decorative, leading one to expect maybe an abstract court-martial. Foreman's ads convey this same grimness and suggestion of horror, and I approached his work with low hopes. *Evidence* surprised me with its amiability, funky Gemütlichkeit, deadpan wit, and personal directness. It's indirect too, out front, as a given of the style, and the night I was there roughly half the audience

Originally published in "Theatre Journal," *Village Voice,* May 4, 1972, 71.

trickled away during the performance, having failed to tune in. But I found it consistently interesting, charming, and enjoyable.

Often it is almost a play. Max is at home alone, trying in a very spaced-out fashion to write a play. This is the play he is writing. Ben comes over, is thrown in through the window. They refer to Rhoda. The speeches are broken up, dialogue all but obliterated. Slow. Incongruities. Metronomes, a blinking light, a drum, a buzzer, quacks to set off time. A beautiful woman downstage petrified, pointing. Later a picture falls off the wall and engulfs Ben. Objects are shown to the audience. Music. Acts and objects come from real life but are transformed: an art not of juxtaposition, like collage, but of transformation. Long looks at the audience. More cords. Lamps. Tilting tables. Cigars. Cardboard boats. Meticulous craft, no illusion.

Here are a couple of quotes from Foreman's notebook (the script):

"1) Basically—audience *waits* for something to happen.

"2) Find a way to make each moment a denial of what is (let something go in your head).

"Finding it increasingly difficult to think of art as giving a positive thing: rather as denying everything so *unknown* can emerge: the not predicted . . ."

He calls it a "tightly-packed sea of void-lets."

But these are the art thoughts behind the work, and aside from their content they have a cold, negative character that is misleading. Ontologically Foreman's work is another story: full of feeling-content and constantly surprising. Never settling down for long into any fictional reality, it runs the danger of turning into art about art, but turns out instead to be a documentary of the artist's head in time. Foreman has developed a personal idiom of persuasive integrity, a distinctive and effective craft. Whether you can dig it probably comes down to whether you respond to Foreman's person as here made manifest. There he is.

The actors are excellently self-possessed and convincing in this intensely disciplined style: Jessica Harper, Dempster Leech, Kate Manheim, George McGrath, and Robert Schlee.

Richard Schechner

on *Rhoda in Potatoland*

A truly fundamental misunderstanding casts Richard Foreman's theater as abstract, surrealistic, dreamlike. Knowing his work from *Sophia = Wisdom* through *Rhoda in Potatoland,* I know that his work is autobiographical, ironic, and *naturalistic* (a word I will explain later). That is, Foreman gives his audience a clear vision of the world as it is, not a distortion of events or images; he provides not a plot in the formalist tradition of Western rising-and-falling action, but in the more naturalistic tradition of the soap opera: a weave of events and reactions among a steady cast of characters portrayed by the same performers over a long period of time.

Foreman's roots—and it's important to locate his tradition (just as Robert Wilson has behind him Wagner, not Breton)—go to the Sartre of *Nausea* and then to Wittgenstein and Heidegger. In literature Foreman is tied to Strindberg, Zola, and Flaubert. The French love Foreman; but only a German knows he's trying to be simple.

What is the basic plot of Foreman's soap opera? It is of a man surrounded by a woman—either a single person shown in many aspects by many performers, or many women all with similar traits—who makes demands first on his attention, then on his energy, and ultimately on his existence. This woman—played by Kate Manheim—moves more and more to the center of Foreman's theatrical world, where she dominates the action, not by scheming or intriguing but simply by being: by existing in a body that is different than a male body.

The sexual struggle in Foreman's plays—as bitter and ambivalent as in any of Strindberg's—is the male's illusion that he can transform the

Originally published as "If Heidegger Wrote Soaps He'd Be Richard Foreman," in *Village Voice,* February 23, 1976, 124.

woman into something like himself by imposing on her the strictures of literature. But she resists "reading" in all of its forms and emerges always naked, and aggressive in her nakedness. In *Pain(T)* Foreman tried to shove words up Rhoda's (naked) ass, but they wouldn't fit—they don't fit: the contradiction is made painfully plain, and it is a very unpopular and difficult contradiction to even say these days: the female body does not accommodate the traditional (intellectual) work of the "artist," especially the writer.

But underneath this nasty sexual theme is another even more fundamental: Foreman sees the world, as Sartre did, as a collection of things tied together and compared. Foreman does not see systems as inherent in the world but only as imposed on the world by human consciousness. The strings in his plays tie things together, activate the space between things, while never denying that this space is part of the way things are. There is no reconciliation in space in Foreman's world. Only the tunnel-visioned depth, the rising and falling stage, suggesting a sea—and how often (as in *Potatoland*) are ships seen nearly falling off the horizon, as if the world could spill out of consciousness and vanish—just as Roquentin was seized with the thingness of the roots of a tree, and nauseated by the tree's utter particularity.

If Rhoda is the *living thing,* animated object, and finally the human being, the men in Foreman's plays are drones—sycophants, extras, mere items. As the years go on even Max, Foreman's alter ego, gets pushed out of the drama. In *Potatoland* Max is upstaged by Foreman himself, who enters the action, steps on stage in person, near the end. This suggests to me a deep change in Foreman's style—I think the series which began at least five years ago is now ending, and in his next works we will see something entirely different.

Let me get back to the basic plot, which is not a story of people so much as it is a set of comparisons, a progression of relationships. The strings connecting things and separating things tie things to things and spaces to spaces in exactly the same way as a line holds a ship at dock. And the progression of comparisons which is the plot of Foreman's plays (counterbalanced by the irony and sarcasm such comparisons stir in Foreman himself) are all about one thing: Rhoda's body and the space she / it inhabits, or literally, takes up. The conflict is Max / Foreman's inability to either maintain or abolish the distance between the "she" of Rhoda and the "it" of Rhoda. In other words, there is the ongoing tension between personality and itness.

Rhoda's personality, in fact, has been coming out for some time—

and it makes itself felt very strongly in *Potatoland*. Previously, only Kate Manheim's eyes showed any expression separate from that which was awarded her by Foreman's direction. But in this production she speaks with intonations and rhythms that smack of "interpretation," of presenting not only attitudes and situations but emotional reactions.

Foreman's theme has always been the main theme, in the West, of the male's attempt, so successful for so long but now breaking down, of denying personhood to women. In Foreman's work this tension animates all the images, actions, and relationships. His world, like Strindberg's, is totally sexualized: the potatoes pushing to get through doors and windows too small to allow entry; the long narrow field of vision with forced perspective; the dancers, always women, staring at the audience, forcing each spectator into acknowledging the role of looker; the framing of slices and planes of actions in an effort to isolate and control ("analyze") them; the ironic or authoritarian voice of the director shouting "cue" or making comments on the action—but always on tape, mediated and removed from the activity going on now; the buzzer-bell-noise-oompah music of the sports event, the combat-as-play.

Potatoland ends with a new image: Rhoda is posed by Foreman in a typical Foreman pose—her naked body arranged on a black tongue of cloth extending diagonally from upstage toward the table where Foreman sits surrounded by his control board (the pilot-manager keeping all things "in order"). But now Foreman, identified because he entered the stage earlier, sits with the Sword of Damocles dangling over his bald head, the actual sharp point of the knife but three inches from his vulnerable pate, the knife's edge unsheathed and sharp: although as director / writer he can tell Rhoda / Manheim what to do, she, as performer, is more and more making her own part. His hegemony is ending.

How personal, how autobiographical, is all this? Without much research there's no way of knowing. But I think it is probably as autobiographical as any work we know, and it is the "true story" (again in the soap opera tradition) of Richard Foreman and Kate Manheim, as seen, thus far, by Foreman. But now Manheim is making her move, as a person, and that will change it all. This kind of world falls well within the range of naturalism defined by Zola nearly a century ago: "I am waiting for a dramatic work which, purged of declamation and free of fine language and pretty sentiments, has the high morality of truth and the terrible lesson of sincere inquiry. In short, I am waiting for the evolution which has occurred in the novel to be accomplished in the theatre, waiting for the day when they both go back to the same spring of science and modern art, to the study of nature, to human anatomy, to the depiction of life in an exact transcript, which will be all the more original and powerful in that no one has yet dared to attempt it on the stage." I think Foreman offers a transcript of what he sees, no more nor less than that. It is enough.

Bonnie Marranca

on *Pandering to the Masses*

Richard Foreman is a philosopher-playwright, an anomaly in the American theater which has never been a philosophical one. In his work with the Ontological-Hysteric Theater Foreman takes as his point of departure the philosophical, psychological, and aesthetic writings of modern thinkers—in short, the Western epistemological tradition. Here is an avant-gardist who is also a classicist. Foreman both challenges and respects the foundations of contemporary thought.

The Ontological-Hysteric Theater dramatizes thinking processes in a highly complex series of images. In *Pandering to the Masses: A Misrepresentation* Foreman creates a reality that reflects his own being-in-the-world, demonstrating in the process a rigorous, alternative manner of focusing on familiar, everyday events. *Pandering,* then, functions on two levels. The subjective nature of the play co-exists with the objective relationship of the audience to the theater event (the central focus of the production).

Pandering appears, on the surface, to have dialogue. However, it is not dialogue as understood in usual theater terms. In Foreman's conception of dramaturgy the spoken language is not only nondiscursive but flattened out through the elimination of inflectional patterns. This flatness is duplicated in the performances of the actors, whose attention to detail and emphasis on natural rhythms of movement and speech produce an extreme naturalism. Speech is disconnected from the speaker by means of interruptive devices such as the tape-recorded Voice (of Foreman) and the voices of the other actors (live and on

Originally published as "Richard Foreman: The Ontological-Hysteric Theater," in Marranca, *Theatre of Images* (New York: Drama Book Specialists, 1977), 3–11.

tape). Instead of engaging in conversational dialogue with one another, the actors, who function as "speakers," serve as the media of Foreman's ideas; they are "demonstrators."

The Ontological-Hysteric Theater is a theater of illustrations in which pictures, continually interrelating with words, replace dialogue. Language exists in the domain of the phenomenological, used merely to indicate a reality in space; space becomes semantic. Foreman is Husserl's "meditating phenomenologist" who, in *Pandering*, meditates on his own and others' attitudes toward art.

In *Pandering* Foreman's focal point is the dual subject of the creation of art and the audience's perception of it. He challenges the popular notion of the acquisition of knowledge about an art object in a dialectical framework that is highly personal.

In performance the actors function cubistically, as multiple facets of Foreman's personality, varying degrees of his subconscious. They also reflect his observations while writing *Pandering*. This accounts for the documentary aspect of the play which, on one level, is a record of Foreman's thoughts while he was in the actual process of creating the play.

The actors, then, serve as blank faces (negative physiognomies) on which Foreman sketches aspects of his "being-in-the-world"; they are representatives of figures of his inner life, playing out the contradictions of his life as a social being. The writer Max is the pivotal figure of the play; he embodies Foreman, the creative artist. Rhoda is the thematic representative of sexuality. Together they manifest the interplay of the intellectual and the sensual which dominates the play. In this display of first-person consciousness Foreman offers the purest form of psychodrama viewed up to this time in the American theater. One scene, in particular, illustrates the personal factor of *Pandering*. Toward the end of the production a man on a bicycle pedals furiously (he represents the energy force of Knowledge) while firing shots at Rhoda. The Voice offers a word—"IKON"—to explain the psychological maneuver which follows this scene. Then, the Voice continues:

> He [Max] inhabits that word. That means to celebrate finally he thinks about his face as being her face so he thinks about his person as being her person finally, and worships it finally, and reads it finally like a wonderful book.

Rhoda is a substitute for Max who, in actuality, is Richard Foreman. Rhoda functions as both icon and idea.

On a second level, Foreman carries on a dialogue with the history of

Western thought in which he attacks conventional modes of acquiring knowledge; in particular, knowledge gained in the perception of an event. *Pandering* is as much a play about Foreman as it is about the audience. In the staging of it Foreman sits in the first row of audience bleachers. From this vantage point he operates the tape system for the production (with himself as the taped Voice that dominates the work) while also identifying himself as an audience member by his presence among the spectators. From this dual perspective of author-spectator the Voice on tape comments on the kinds of responses elicited by the "old theater." At one point in the production the Voice declares:

> The old theater would prove to you that Max is dancing the way that he is dancing, by which is meant, his motives, proven real and genuine, and you are convinced in a way appropriate to the theater.

Interestingly, *Pandering* bears a striking resemblance to Peter Handke's *Sprechstücke,* which eliminate dramatic dialogue, employ "speakers" or "demonstrators" rather than characters, construct a dialogue between stage and audience, and debate with conventional theater. Likewise, Foreman's work can be looked upon as "autonomous prologues to old plays" (Nandke's phrase). And *Pandering to the Masses: A Misrepresentation* is as ironic a title as Handke's *Offending the Audience*—both are the "speak-ins" *(Sprechstücke)* of their authors.

Pandering is presented to the audience in the form of a "lecture-demonstration." Verbal and visual images accompany Foreman's running commentary. For example, the Voice remarks on the occasion of one of Rhoda's adventures:

> every experience, though perhaps peripheral to the primary revelation of knowledge, friendship, and inventiveness . . . can be a learning experience if allowed to take its place in the mapping process of one's private adventure and spatial self orientation.

Projected slides ("legends"), which one may consider a more sophisticated variation of flash cards than those used by teachers in classrooms, carry content, which corresponds to or contradicts the image on stage, or describe an image that has already appeared or is about to appear ("He goes to the wall / finds two peepholes"). At the opening of the piece the Voice virtually insults the audience's intelligence in a "lecture" that declares:

> You understand nothing. Max regretfully concludes that you who watch and wait have unfortunately proven through your actions and reactions that cer-

tain subtle, exact, specific and necessary areas of understanding are not available to your—

The audience is left to fill in the blank.

Frequent "recapitulations" in the text reflect Foreman's didacticism. Yet they also have other functions. They serve as a flashback technique, or to further the action of the production by the interaction between Foreman and the actor on stage; often, they reinforce the memory of past events or clarify certain points of the text. In one instance, a "recapitulation" is calculated to make the audience reflect on the associative mode of perception. Foreman chides the audience, "Do you think using the associative method? Everyone does, you know." Like Gertrude Stein, whose writings, he has admitted, have been a major influence on his aesthetic theories, Foreman contrives to destroy associational emotion in the experience of a work of art. More significant, however, is Foreman's ability to create in his work what Stein referred to as the "actual present." *Pandering* exists in a dual framework: as the actual diary —the personal notes—of the playwright while he was writing the play over a certain period of time, and as a complete play unfolding in performance time. The past and present merge in the actuality of the performance.

In another allusion to a major Steinian concern, the Voice directly confronts the issue of which image on stage takes precedence over another in the sequence of events:

> You can either watch Max writing it, or you can watch what he is writing. But you can only watch what he is writing after he is writing it, and in that case your expectations are in a different direction, are they not?

Similarly, in a lecture entitled "Plays" (1934) Gertrude Stein observed that the audience is always ahead of or behind a play on the stage, never exactly "with it."[1]

To both Stein and Foreman it is the *conscious* act of experiencing events at a certain time and place that is important. In *Pandering* Foreman seeks the triumph of the conscious over the unconscious.

Pandering is a consciousness-raising piece—a teaching play—whose goal is to make audience members aware of their moment-by-moment existence in the theater. For this reason alienating devices obtrude throughout the production. Foreman continually breaks down the pro-

---

1. *Gertrude Stein: Writings and Lectures 1909–1945*, ed. Patricia Meyerowitz (Baltimore, Md.: Penguin Books, 1971), 59.

duction into smaller and smaller units or frames. He imposes a play, *Fear,* within the play. Many of the actors' lines are prerecorded on tape in Foreman's Voice, which interrupts them; the actors interrupt each other's speeches when each word in a sentence is spoken by a different person. Buzzers, loud thuds, and music focused directionally by four stereo speakers which surround the audience punctuate the actors' words. Scene titles, when they are used, break the flow of the production.

Foreman's use of these "alienation effects," more directly, his conception of a play as a "teaching play," reflects his Brechtianism. However, he is formalistically more radical than Brecht. Foreman breaks up his scenes into smaller and smaller units; Brecht divided his epic structures into unified scenic elements. Foreman is a minimalist concerned with instantaneous perception; Brecht's view was epic and historical, concerned not so much with momentary perception but with critical thinking which would lead to political activism outside the theater. In Brecht's productions the actor "commented" on or "quoted" a past action; *Pandering* strives to create a continuous present even as it treats events of the past, that is, Foreman's thoughts while writing it. Furthermore, though both artists devised dialectical theaters of illustration, Foreman has gone further than Brecht by moving the dialogue from its fixed position in a play on stage to a dialogue (metaphorical) between stage and audience. In this way, the dialectical aspect does not remain solely in the framework of the play on stage but occurs directly in the relationship of the audience to the production *in process.* (Handke has also accomplished this in his *Sprechstücke.*) Foreman always demonstrates *how* the play works.

Pandering emphasizes its producedness, that is, the interconnection of its parts. This focus on structure necessarily compels the audience to scan it for minute alterations. By calling attention to itself—how it works—it stimulates the audience's powers of perception. The Ontological-Hysteric Theater is radically opposed to the traditional theater (what Brecht called the "culinary" theater), which feeds information to the audience by suggesting the "proper" emotions and responses to stage events. In Foreman's theater there are no touchstones, no recognizable pegs on which to hang conditioned responses or ideas. The world of the play, while not duplicating reality, suggests a way to view life in the real world. Through Foreman's reduction of his personal life to a series of images one is led to perceive things as they are in themselves—not by learned patterns of perception but in an unconditioned way. Observation supersedes memory.

Foreman's emphasis on changes, conscious response, the present moment, his use of "recapitulations" and taped directives for viewing *Pandering,* force the audience to be aware of the making-of-theater. Thus, the process of *Pandering* is always evident. During performances the actors virtually construct the set of the play, redistribute space (expanding, contracting, deepening it) by means of props, sliding frames, and drapes. The actors themselves are sometimes "constructed." Fitted at times with objects and cloth, they appear like assemblage art. In one scene dolls are strapped to the legs of the actresses for a musical number.

In the world of *Pandering,* where gesture is dissociated from language or merges with it, where language is fragmented and thought dislocated, time is experienced in terms of changing spatial relationships. In the design of the set Foreman plays purposefully with perspective: A road leading to a house at the back of the playing area narrows to its end ("Try looking through the wrong end of the telescope. Everything looks sharper, doesn't it?" asks the inquisitive Voice). Many scenes are presented in slow motion, suggesting a two-dimensional, painterly perspective. The actors frequently stare at the audience or gaze sideways; other scenes are presented from the perspective of the picture-frame stage. The continual rearticulation of space which Foreman's long and deep but narrow loft theater affords complicates perception of movement and disorients the audience which must accordingly change its field of vision to accommodate the variety of spatial configurations.

Foreman's idiosyncratic use of strings, which dangle from the ceiling and stretch in horizontal rows or diagonal crosses about the performing space, sectionalizes space and cuts it into geometric shapes. Another use of the strings has more to do with Foreman's insistent directorial focus on elements within the stage picture: his pointing out certain correspondences between the words and the images. When a performer, for example, draws a string from one end of the space to another until it touches a person or object, that person or object is defined in an exact point of time and space, as well as in reference to other activities on stage. In this way the world of *Pandering* presents a diagrammatic reality whose system of reference is entirely within the play as performed. Foreman's work is conceptual art, that is, self-defining.

The rhythmic element of the piece is carried from unit to unit. The sound of a metronome during the production articulates its beats, affirming the musicality of the work. Foreman's work in the theater parallels the trance or minimal music of such diverse composers as Philip Glass, LaMonte Young, Terry Riley and Steve Reich, in whose compositions the accretion of sound is a key structural feature. Glass, in particular, has written:

> nothing happens in the usual sense . . . instead, the gradual accretion of musical material can and does serve as the basis of the listener's attention . . . neither memory nor anticipation . . . have a place in sustaining the texture, quality or reality of the musical experience.[2]

His statement accurately describes the ambience of *Pandering,* which is constructed from the gradual build-up of small units of composition.

Conversely, subtraction—pause or silence—is important to *Pandering* as an interruptive device and a way of slowing down the performance, as is playing the taped Voice against the natural voices of the actors by subtracting them from their words. John Cage was the first to regard silence as a viable structural unit of music, overthrowing much of traditional compositional theory which had viewed time in music as an empty unit to be filled. In the dance world Yvonne Rainer and others of the Judson Dance Theater and post-Judson period experimented with time as duration and the subtraction of movement. They were often joined by artists such as Robert Rauschenberg and Robert Morris who worked very specifically with new concepts of movement through time and the placement of objects in space. Foreman has solid foundations in post-Cagean aesthetics as they have filtered through the worlds of art, music, and dance. His exploration of movement and spatial organization, elasticization of time, and radical situation of objects in the construction of his production have been, and still are, dominant preoccupations of New York avant-garde artists.

Foreman's application of silence is significantly demonstrated in his use of tableau—the subtraction of the moving image. Tableau is *Pandering*'s chief unit of composition, a still life which frames the action and "quotes" it. This quoting of gesture is another Brechtian technique that finds expression in Foreman's theater; however, Foreman's employment of tableau, because it occurs more frequently and lasts longer than Brechtian tableau, elasticizes time as well as continually disrupting the spatial flow of the production. In tableau Foreman's actors appear as frozen voids in space—like the chalk-white faceless forms in a Chirico landscape—until they are revived by the tape machine or a change of scenery.

2. Program Note to a concert at Town Hall on May 6, 1973, at which Philip Glass and his ensemble performed "Another Look at Harmony, Parts 1 and 2."

Tableau is used in various ways: the duplication of gestures in foreground and background; close-up perspective; the inclusion of a single kinetic element in an otherwise frozen picture; a confrontation of the audience vis-à-vis the actors in a frontal position; and for iconographic effect. Finally, by employing tableaux which throw certain elements of the production into high relief, Foreman is able to bracket perception of events on stage, thereby drawing attention to particular elements of the stage picture.

The framing of events on stage is paralleled by the framing of objects in which a single element is presented in close-up against a larger, more complicated background of activity. For instance, in the palm of his hand an actor in a foreground position holds a letter in an envelope which is affixed to a plate: It is another way in which Foreman breaks the continuity of composition by extreme reduction of a scene or gesture. Windows, boxes and cutouts in the design of the set frame objects or people, such as Rhoda's breasts exhibited in a door frame.

Curiously, Foreman's published notes and manifestos reiterate his fascination with framing. Sentences are often subdivided by parentheses, brackets, and equation symbols; rather than flowing smoothly they focus on single elements.

The framing devices and the tableaux of *Pandering,* in addition to disrupting the flow of time, draw attention to its passage. Time exists, as it were, in the continuous present of the dream world where images of the subconscious appear, drift away and then reappear, or collide with other images. Space repeatedly changes its contours in defiance of physical laws so that a wall of a room gives way to a jungle and scenes shift easily from outdoors to indoors. The people who inhabit this surreal world are free to roam with abandon through a series of adventures that take them back and forth in time and place. It is the landscape of Foreman's mind, the image-activated visions of his subconscious. Scattered about in this world are fruit, an oversized horse and giant pencils, croquet balls, stuffed animals, a pistol, snake, and bicycle—all of which have symbolic value in Foreman's psychodrama. These are the symbols of childhood, of violence, of power and fear, temptation and sensuality.

In his situation of objects and people in *Pandering* Foreman recalls similar styles of the personal, surrealist films of Jean Cocteau and the American experimentalist Maya Deren. Foreman shares with them his love of melodrama, eroticism of violence, the placement of the human figure against a plane, narcissism, and the distortion of time and space through the use of mirrors and walls. However, Foreman's work differs from their classical surrealism in its demand for rigorous control rather than spontaneous expression, the importance given the phenomenological activity, and its high degree of cerebralism. Seen in another light, Foreman's filmic inheritance may be what P. Adams Sitney observed as the American avant-garde cinema's unacknowledged aspiration: "The cinematic reproduction of the human mind."[3] *Pandering to the Masses: A Misrepresentation* is a *theatrical* reproduction of the human mind—Richard Foreman's. He creates in his theater a new way of thinking about the theater event, a greater consciousness of art. Not only does he make theater going more meaningful, but life as well.

3. P. Adams Sitney, *Visionary Film: The American Avant-Garde* (N.Y.: Oxford Univ. Press, 1974), 408.

Frank Rich on *Penguin Touquet*

Richard Foreman, the director, doesn't dream big dreams. He dreams mesmerizing small ones and then inflates them until they fill the stage. In *Penguin Touquet,* his new theatrical piece, Mr. Foreman is dreaming about spinning waiters, bearded saxophone players, blind men, large gray rocks, scissor-crazed barbers, fish soup, a town called Great Poetry—and, yes, penguins. Are his dreams any more profound than yours or mine? No, but then you and I are not capable of turning our subconscious ramblings into an evening of hallucinatory theater. Mr. Foreman is. *Penguin Touquet* may not be an intellectual revelation, but it's surely one of the season's most audacious displays of fantastic stagecraft.

Although this production unfolds in the Public Theater's spacious Newman Theater, it doesn't much depart from Mr. Foreman's past antics at his own tiny Ontological-Hysteric playhouse. Once more the proscenium is bisected by string; the leading lady is still the alluring Kate Manheim. In *Penguin Touquet,* which runs a refreshingly taut 80 minutes, there is also a story of sorts. Miss Manheim plays Agatha, a latter-day Dorothy who searches for an ontological Oz. The wizard is a psychiatrist (David Warrilow), who may or may not have the power to lead her into a "different form of life."

Well, let's not be literal-minded. What counts most here is Mr. Foreman's gift for shaping actors, scenery, sound, lighting, and architectural space into a cascade of animated images that have their own internal flow and logic, their own cause and effect. Some of the images are funny or beautiful or frightening, and many are haunting. In *Penguin*

Originally published as "Stage: 'Penguin Touquet': Hallucinatory Fun," in *New York Times,* February 2, 1981, C14.

Touquet, Mr. Foreman has managed to fold the spooky visual styles of Magritte and de Chirico, as well as Brechtian theatrical technique, into a show that occasionally recalls, of all things, that old psychoanalytical musical, *Lady in the Dark.*

The action unfolds in a mysterious funhouse of a set, designed by Mr. Foreman and Heidi Landesman, which looks like a cross between a giant eye chart, a Left Bank Paris bistro, and a chessboard. It is decorated with incomplete words ("differen," "resoulti"), and its components are always on the move. The stage frame is bordered by lights of all kinds, which sometimes blind the audience. The sounds include snippets of movie music, pounding heartbeats, rhythmic chants ("I am on fire, am I not!"), loud breathing, and deafening electronic blasts. Mr. Foreman long ago went on record to the effect that the artistic experience must be an ordeal.

Yet *Penguin Touquet* is not painful, and it's rarely boring. The best vignettes really do have the lucidity and power of dreams. In one of them, Agatha pounds a piano keyboard with one of the ubiquitous rocks, only to be told by a godlike voice that "The rock can't make a career of playing the piano" and that it will soon be "pulverized, mixed with other things and turned into a road." The indominatable Agatha asks, "Where will the road lead?"—but she gets no answer. Instead, the cast breaks into spastic gestures that suggest some mass electrocution of the psyche.

I was also fond of a sequence in which Agatha starts to whistle a symphony, ends up whistling a popular tune and soon gets swept away in a frenzied tap dance. Somewhat later the penguins appear—they may or may not symbolize sexual terror—and the lighting (superbly designed by Pat Collins) dims into a spooky twilight that promises a carnal hurricane. Another motif of *Penguin Touquet* is gold. Disgusted to find some nuggets in his soup, Mr. Warrilow plaintively asks, "Would you mind if I broke down and cried?"

As always, Mr. Foreman plays manic tricks with time. The actors' movements frequently speed up or slow down, silent-movie style, and the one poor soul who attempts to wear a watch gets his wrist burned. The evening's choreography is unfailingly precise; the associative links between segments are so clear that we never lose our way in the director's iconographic map. We're ready for anything: heads of cabbage, chandeliers, a huge snowman. When characters are struck by nosebleeds, we accept the attacks as Mr. Foreman's equivalents to a conventional play's emotional catharses. Indeed, when the bleeding Agatha retreats from her self-discoveries at the end, we're even moved. Her final, questioning refrain—"When you see something out of the corner of your eye, do you really see it?"—perfectly expresses the melancholy of a dreamer awakened.

Though acting is not required, the performers are an integral part of the fun. Miss Manheim is a splendid girl-woman, capable of both primal rages and sexy flights of daffiness. The spindly Mr. Warrilow, dressed like a sepulchral floorwalker of the soul, is a grave deadpan comedian who can break into funny paroxysms of mock-vomiting. Gretel Cummings is quite amusing as a "grande dame" who erupts orgasmically when she walks across the stage with an old-fashioned radio between her ankles.

And what does it all mean? Mr. Foreman provides the audience with a combination glossary-plot summary that explains his symbols and recaps his concern with such philosophical polarities as mind and body, nature and culture, feeling and thought, and so on. Be assured that there's little new or surprising in any of it. But one doesn't attend this show to discover the meaning of life, or even the meaning of the title, *Penguin Touquet*. One goes instead to watch Mr. Foreman transform his idiosyncratic stream of consciousness into a flood of theatrical magic.

Gerald Rabkin on *Don Juan*

In his Ontological-Hysteric Theater of the 1970s, Richard Foreman created a microcosm in which words were treacherous. They formed declarative sentences which in themselves made logical sense: they described, informed, generalized, cautioned, exhorted—but they did not combine to create the verbal construct we call a "play." Each sentence—occasionally each word—existed not in relationship to conventional linguistic meaning, but as one movement in the shifting intersections of lines of theatrical force which formed alogical theatrical images composed as much for the eye as the ear. "Listening and looking," he affirmed in *Lines of Vision,* "become interchangeable."

After the closing of this theater in 1979, Foreman did not abandon his re-vision, but has shown himself increasingly willing to focus it on verbal texts other than those he has himself created. In 1977, at the behest of Joe Papp, he had directed a Lincoln Center production of *Threepenny Opera* which, though unconventional, hewed to the letter of Brecht and Weill, but it seemed a deflection of his theatrical energies, which were still concentrated then in his theater on lower Broadway. Now he is no longer as firmly wedded to the role which accorded him prominence in the seventies—the playwright, director, designer, impresario, who distills into himself all specialized theatrical functions. His recent work on view in New York—productions of Botho Strauss's *Three Acts of Recognition* and Molière's *Don Juan*—reveal his assumption of a more traditional directorial role, as interpreter of contemporary and classic dramatic texts.

Traditionally but with a difference: Foreman remains committed to the ideas which inform his Ontological-Hysteric work. Language is still

Originally published in *Performing Arts Journal* 18 (1982): 67–70.

treacherous—indeed, even more so when it is encrusted with the sedimentation of history. Remember, of all American theater experimentalists, Foreman is the one most consciously indebted to new European—structuralist and post-structuralist—modes of discourse. With Barthes, Derrida, Foucault, et al., he rejects the realist or authoritarian heresy that the critic (or director) can make definitive contact with some ultimate, residual meaning when, in reality, he is simply transcribing a code—or a series of interlocking codes which can be deciphered but never fully recovered. In his production notes for the Guthrie Theatre *Don Juan*—which, with some revision and recasting, was essentially the production transferred to the Delacorte this past summer—Foreman affirms his discipleship: "Following the lead of contemporary French theorists, I find it most productive and illuminating to regard the written text of the playwright as the 'deposit,' the 'tracings' of what obsessed him as an individual."

So the theatrical problems that Foreman-the-interpreter must confront are not simpler but more complex than those faced by Foreman-the-*auteur*. In his Ontological-Hysteric work, the "intertextuality" confronted is contemporary, rooted in his singular consciousness—an intertextuality of linguistic and gestural discourse rooted (in the language of contemporary French theory) in a decentered present. Interpreting Strauss obviously presents fewer problems in this regard than pursuing Molière through the maze of history; Foreman's sensibility is clearly attuned to the Brechtian-influenced texts of the contemporary German theater (his own debt to Brecht is profound). But Molière's *Don Juan*? An acknowledged "classic" text? If the authority of a univocal reading (assumed whenever it is demanded that the text *speak for itself*) is rejected, how is the director to physicalize the intertextuality of past and present? How can the "tracings" and "deposits" of the past be best uncovered? As Elinor Fuchs pointed out in the *Village Voice,* it is a problem Foreman shares with a generation of "postmodern" directors— Lee Breuer, Joseph Chaikin, Andrei Serban—as they increasingly accept the challenge of confronting "classic" texts.

Aesthetic problems are compounded by production exigencies. Obviously, a performance in the Delacorte's sylvan setting breeds traditional theatrical expectations. Though it is doubtlessly courageous of Joe Papp to challenge his summer audience's orthodoxies with radical productions by such as Foreman and Breuer (whose pop *Tempest* summer before last infuriated many), we can assume that he would not assent to a wholesale "deconstruction" of the literary text—in the manner, say, of Grotowski's *Faustus* or Serban's *Trojan Women*—in which the classic text is dissected and recomposed. Experiment has its limits. The model becomes Brook's *Midsummer*: an innovative performance text superimposed upon a faithful rendering of the verbal text. That Breuer and Foreman have been willing to accept this model and the challenge of an audience unschooled in experimental theater conventions is to be respected. But in testing unfamiliar waters they are thrust into tides they cannot fully control.

Foreman's choice of *Don Juan* affirms his post-structuralist concerns: for it is a text which embodies a powerful myth which has been variously interpreted from the 16th century to the present. As Shaw writes in the preface to *his* Don Juan play: "The lesson intended by an author is hardly ever the lesson the world chooses to learn . . . Don Juan became such a pet that the world could not bear his damnation." And Molière's version of the myth, violently attacked and withdrawn after its initial production, has itself been subjected to wildly divergent interpretation. For Foreman, the attraction of the play lies in its historical displacement: it harkens back to Tirso de Molina's Enemy of God and beckons forward toward the rationalist Enlightenment (2 plus 2 equals 4 as the ultimate truth). So in his production Foreman aims not at a univocal reading of the play or the myth, but at a synthesis of opposed values: "a series of articulated echoes and reflections as the characters mimic both each others' and their own language and gestures" [Guthrie notes]. Affirming Derrida's deconstruction, he aims to show the text to be woven from different strands which can never result in a synthesis, but continually displace one another. For example, Don Juan is at times a hypocrite; but in being so he echoes and magnifies the hypocritical world around him. He is, therefore, greater, more intelligent, more aware than those who surround him—even to the extent that he is morally worse.

So Foreman's directorial strategy is to corporealize these "echoes" and "reflections." He invents a chorus of chalk-faced aristocrats who serve a double function: they represent the repressive hypocrisy of Don Juan's era, and, as contemporary actors, distance the performance by serving as a buffer between the contemporary audience and the "play proper." And at the shifting center of the performance Foreman encases John Seitz' Don Juan in a rigidly choreographed style derived from baroque opera, to modern eyes a style crab-like, evasive, imprisoningly formal. Other echoes and reflections are evoked by visual strategies derived from the Ontological-Hysteric vocabulary: curtains rising and falling on changing tableaux, quartz lights blinding the audience, mirrors literally reflecting its visage, strings—lines of force—bisecting the stage, fragments of lost linguistic codes lettered on bulbous black sconces, bizarre bursts of disorienting sound and music.

Space does not permit a detailed performance critique, but it is clear that the given compromise at the heart of the production continually shifts the level of theatrical discourse. No other performer matches Seitz' intense stylization; most attempt traditional performances as befits their training. We recall that Foreman consciously avoided the use of professional actors in his O-H work in pursuit of the deconstruction of the concept of "character." Here he is caught in an insoluble dilemma that envelops both dramatic and performance texts. In accepting the textual immutability of the play his revisionary strategies must all be embodied in his *mise en scène* and hence risk the dangers of imposition and redundancy, dangers plaguing all such experimental versions of "fixed" classics. And working with traditional actors often means imposing on them values and techniques most find inauthentic.

Yet surely this attempt to bridge conflicting modes of theater discourse must be valued for its intelligence and aspiration, if mixed achievement. Without such ecumenical gestures our theater will be trapped in a Tower of Babel where each cannot understand each, except in the lingua franca of banality.

Gordon Rogoff on *The Cure*

Hovering like a warning or an admonition over the tiny space in which Richard Foreman's *The Cure* moves through its subdued meditations is a sign saying No Secrets. As usual with Foreman, the words don't slip easily into a category or conclusion: most of the time, he has set up oppositions and unanswerable questions which act as a dialogue with the self. This time, however, with No Secrets hanging over the proceedings, he does something new and ironic—a confrontation of the most private theater artist we have with the impulse to tell everything he knows and feels.

It isn't probable that Foreman's privacy can be breached even by Foreman himself. That he's theatrical is, I suspect, his personal salvation: using a voice and iconography all his own, he releases himself to open struggle, a continual play of words and energies that probably make his bad dreams bearable. In *The Cure,* what is seen and heard most certainly must be all we need to know. While there may be a temptation to cast interpretative veils over his signals and images, it is sufficient to take the journey as given—no more, no less than what it looks like or seems to be at any passing moment.

Designed by Foreman, the set looks like a fun-house funeral parlor. Gray curtains with black lace trim hang over the bottom half of the side and back walls. Most of the objects and props are various shades of black and gray, all of them surrounding or placed on two tacky pseudo-oriental rugs. A black box with unlit candles on its four corners await the actor who will climb into it for a time. What appear to be black-curtained election booths stand on either side of the stage. On upper

Originally published as "Richard Foreman: Internal Gestures," in *Village Voice,* May 28–June 3, 1986, 81.

shelves in the back are a box of Kellogg's Corn Flakes, some empty soup bowls, and another bowl of fruit. On a shelf above them, a curved scimitar sits impassively as all the other objects, another thing to be used, even if that use can never be palpably clear. A dark thronelike chair sits to one side of the back wall; on its crest is an inlaid photograph of Tchaikovsky. Intruding on all this rational gloom are still more hanging mechanicals—a series of colored discs that are eventually set in circular motion by an unseen hand. Foreman gives his actors a space that is deliberately unsettling and only pseudocomic. Yet one thing is clear: unlike most imagistic theater, *The Cure* is not trying to paint pretty pictures.

The pictures here are almost accidental deferences to the theatrical mode. Foreman's biggest open secret is his passion for words. Placing the three actors—Kate Manheim, David Patrick Kelly, and Jack Coulter—in a hermetically sealed parlor, Foreman gives them nothing to do but enact or present the story of a writer possessed by language and argument. At one moment, as if summoned by a sudden glimpse into ecstatic realms, the men dash and shake their bodies into staccato riffs; at another, they suddenly clap their hands repeatedly on their knees. Kelly is imperturbably cool; Coulter leers and lurches, trapped in his own strenuously induced steam heat. Meanwhile, Foreman spreads a glistening pattern of questions, observations, and half-answers over almost every move and tic. "Can the truth be conveyed in a story?" he has an actor ask, and we know implicitly from the work itself that Foreman finds stories a lie.

Wearing body mikes so that their voices are heard over speakers, the actors are further distanced by the continual accompaniment of alternately ominous and whimsical music, that too scored by Foreman. Can it be an accident that the tunes and rhythms have an effect not unlike the scores for Chaplin's silent films, a kind of tinny, filtered, music-hall quality suggesting paranoid games? Consider, too, that *The Cure* is also the title of one of those films. (With Foreman, some of the fun lies in playing the games.) With one of those old-fashioned radio signals denoting the hour or the announcement of the station, each selection is halted and a new one begins. Foreman's theater is a conspiracy on behalf of words: these orderly, marked divisions are designed to convey us into a listening mood; the text may be densely packed, but Foreman makes it unavoidable.

With po-faced intensity, the actors move from one object to another while speaking their litanies. "Here is an important dream," says Manheim, telling us that she "woke up one morning and found the world was all I desired of it." She doesn't, however, tell us what those desires were. Another section begins with the demand to find out something—"how to punish a man approximately five feet eight-and-a-half inches tall." And again, nobody finds out how to do any of it. In still another section, the actors declare that nobody knows much about them, but repeatedly they conclude that they don't care. Further on, they are implored not to guess or analyze. Warned later that "it's not wise to talk about things that scare you, make you cry, make other people mad at you, or jeopardize your economic security," Manheim replies that she "never does—such things are private." Yet she can't avoid her own subversion: "Sometimes I talk about them without knowing."

Toward the end of this one-hour ceremony of the half-alive, Foreman flirts dangerously with what looks like a conclusive statement: "The cure," we learn, "is in the pain." This is turned quickly, however, into its opposite—"There is no cure." And then again, it becomes another question: "The pain of the cure is the cure?" The danger has passed just as swiftly as it emerged. Foreman is no more likely to provide homilies than any other inquiring artist. If a man thinks he's eating normal cornflakes, someone is bound to tell him that he's wrong. "You'll never solve it," says Manheim.

If he can't solve it, at least he can share some of the pain. Manheim's brittle presence—limping wrists and high heels conveying her over territory that never quite explodes under focused assaults—is the walking-talking emblem of Foreman's quizzical distress. In one of those lines that keeps escaping into his own prepared oblivion, he has her say that she heard the "glacial cracking of her own emotions." Quoting Alfred North Whitehead's "nothing in excess" as "the motto of the philistine," Foreman provides an excess of vigorous speculation that is as glacially cracked as Manheim's elusive emotions. *The Cure* may remain private at the end, but if you can bear its solemn fun, it offers the gift of what theater does better than the other arts these days—what Foreman calls here "an internal gesture of the mind." Glacial, yes, but insistent and deeply felt.

Erika Munk

on *Film Is Evil, Radio Is Good*

At one point in *Film Is Evil, Radio Is Good,* a row of signs lights up; in little letters they announce we're "on the air"; they also look a bit like the exit signs in a movie theater. What they say in big letters is "EGO." Ego, however, offers neither a true voice nor a way out in Richard Foreman's new play. On the contrary: Ego as vanity, ego as the desire to hear one's own voice and see one's own image, ego as the will to control others by controlling voices and images, has thrown our part of the human race—the "developed" world—into a self-reflecting pit where ego as autonomy has been lost.

On a sound stage hung with Hebrew lettering and a multitude of clocks whose hands are close to midnight, or perhaps noon, the invisible delights of radio are pitted against the tyrannical images of film, Echo against Narcissus. Film is coercive, radio is freeing. Film is aggrandizing, radio is faceless. Film is insatiable, radio is soothing. Film makes you sit, radio lets you walk. However mysterious or hilarious a moment is, however many layers of meaning can be drawn from it, this argument is the play's constant text. Foreman's characters are struggling with the various ways human consciousness has been displaced by the machines of expression, and with our manifold violations of the first two commandments, as we turn our all-too-human selves into god and worship our own images.

These are hardly new subjects. Over the last fifteen years or so, serious thinking about "the media" has changed our ideas of oppression, expression, and reality. That art's now more often about artforms than about anything else is a cliché, though true. The very existence of an

Originally published as "Film Is Ego, Radio Is God," in *Village Voice,* May 19, 1987, 91.

94

unmediated (forgive the pun) anything-else has become an open question. In American theater, this question's been dealt with extensively but evasively: Commercial productions embrace the manipulations of commercial television; experimental performance satirizes them (gently) or stages its own seduction by film. For Richard Foreman to make live theater out of *thinking* about media is a grand, paradoxical, absurd, touching project. We may have reached "the end of in-itselfness," but Foreman tells us about it in the most in-itself, fleshly, fallible medium of all, and with this paradox illuminates the grief and pratfalls of people who've lost themselves as a result of their own inventions, and are looking desperately for a way out.

Kate Manheim, her deadpan gaze fixed somewhere slightly over our heads, plays a radio broadcaster named Estelle Merriweather, the most complex and least victimized version so far of the ever-questioning character she's played in Foreman's works for almost twenty years. She appears first as an ominous dark figure crossing the stage; wears technician's white most of the time as she praises radio, fights the camera, and rejects the hero; puts on a high black hoodlike cap—something for a sorcerer or inquisitor—when seeking spiritual illumination. "Did you ever hear me late at night, broadcasting to whoever may be out there in the darkness?" A split spirit, she drifts between technology and mysticism, bossy coolness and a wry yearning that would be lyrical if she didn't catch it at the last minute. "Man has failed and must be elevated to ritual," says Estelle.

This is not just generic "man." "Man" seems to be male. Estelle's admirer Paul Antonelli is performed by David Patrick Kelly as a sweetly serious, confused soul, though in one memorable scene he bursts into lust with two women in a pew. Saints Paul and Anthony, no doubt. Foreman doesn't leave Man at that, however. Man is didactic—"Don't run away until I can find out how to explain everything"—and controlling; Man is in love with film. In *Rhoda in Potatoland,* Foreman left his position at the controlboard and came on stage to argue with Rhoda; now he appears as the star in Babette Mangolte's film-within-the-play. Radio-Rick-in-Heaven rejoices in the camera; Radio-Richard-in-Hell presumably knows better. Foreman, whichever one he embodies, is elevated to ritual, though perhaps not in the way Estelle meant: as the camera fixes on his intelligent spaniel-eyed face, his image turns—without really much changing—more and more threatening, more and more authoritarian. At the film's end Foreman humbles himself to the camera in a Moslem's attitude of prayer, and vanishes.

Of course theater, for all its three-dimensional mortality, is just another medium. The third major character, Helena Sovianavitch, is the very spirit of the histrionic, all flesh and mannerisms, and Lola Pashalinski hams her up marvelously. Every grand gesture and rounded elocution reminds us that live performance is unthreatening because it doesn't try to replicate or replace reality—and that it was a centuries-long opening wedge for media domination, tempting people to perform themselves. When Pashalinski hands us a Chekhovain moment, it's breathtakingly funny, theater's last gasp of dominance. Helena Sovianavitch is the owner of the station, but not for long; it's as doomed as the cherry orchard.

All these interpretations are provisional and maybe flat wrong. Certainly they are much too tidy in light of the event itself. Twelve NYU students play a multitude of roles, including Charity, Purity, Love, and Virtue. A giant egg is handed about and set down here and there. Smarmy moments of radioese come and go. Everyone dances hysterically, freezes suddenly, dances again. The deep stage is crammed with constantly changing props, lights and music hurl us abruptly from one emotional world to another. No matter how much intellectual fantasizing the show prompts, it's enmeshed in sensuous event: exactly the opposite of what happens in a lecture hall, and the great charm of Foreman's theater.

As the play unfolds, the images are more and more Judeo-Christian. This is "a fallen world that worships a multitude of graven images," and a smug little golden calf appears, soon followed by Paul, carrying a side of veal. The students hold placards with Christ-like faces. Helena walks around with a collection plate on an enormously long pole. Estelle is looking for an alternative to film, but can she find it in this dubious religiosity?

Perhaps she'll find it in books. A table's covered with medium-sized handy silver-covered books, decorated with tiny glittering red hearts. Are books the real magic, or countermagic? Are these real books, or totems? No one can read them. Dead texts, it seems, hearts and all. Hearts in particular: "In today's world some people experience a certain deficiency of emotion"—the line gets a big laugh.

"The microphones in this beautiful modern studio have turned to ice" doesn't, and shouldn't. Suddenly, it made me aware that while audiences over the years have found Foreman's work chilly, or at least on an exceedingly high level of abstraction, this play isn't at all deficient in emotion. Something new—though it was visible in *The Cure,* last year—is going on. Not, god forbid, sentimental realist "feeling," but a powerful sense of balked connections, of loss and desire. "I put my hand out to touch the piece of ice he extended to me, and as my hand approached his hand, I suddenly saw it fall into a second dimension of space, through a door that existed in my imagination. As it happened the name written on the back of my hand that day was 'heaven'? And lifting it to my forehead, it was as if I sped through several doors simultaneously. The door of the present and the door of the past. Was I forgetting or was I remembering? It doesn't matter, it doesn't matter." This is not a speech to the mind alone.

The audience knows something the protagonists don't, and the sadness of this knowledge is a final shadow—sometimes ironical, sometimes sad—on the action: the film / radio argument has come too late. The play's oppositions are misleading, because television has taken over. The real poles are the mechanical media and the living self, or perhaps the machine and the spirit. Both radio—think of Hitler's radio speeches—and film started what more advanced technology is completing. There's no turning back. Estelle says, "Here's an idea of my own invention. De-hypnotize the brain," and finally pulls the plug so she can listen to things as they really are. Through earphones.

Estelle searches for a way out in the real, or spiritual, world. Foreman / Paul / Rick / Richard embody, and fight, a self-referentiality so extreme it becomes a form of modesty and universality. Look how trapped in my own head I am! Isn't that silly? Aren't you in the same boat? Foreman's vanishing at the end of the film can be taken as a miraculous loss of self, or a disastrous one. That he isn't present on the stage, that he's no longer visibly controlling every move, is, however, a lovely step in a piece that is both marvelously like Foreman's work before he got famous, and miles beyond both that work and what's come since. A colleague argues that the play is not about technology but about artforms. I can't imagine a work called *Painting Is Evil, Music Is Good* imbued with the same feeling that our lives are at stake.

Kenneth Bernard on *Lava*

Those who did not see Richard Foreman's recent production of *Lava* at the Performing Garage missed an unusual opportunity to see his work through a glass clearly, a philosophic pronunciamento. As his theater's title (Ontological-Hysteric) has always implied, Foreman has always more or less dramatized the anxiety of being. His concern has always been to break through the conventions to a raw feel of reality / truth. In the process he beats epistemology to death and still fails. His theater is the hysterical drama of that failure.

Lava, like other productions, is foregrounded by props (including his essentially miming cast)—the incredible variety, thereness, and fundamental mystery of the things of this world. We are surrounded by them, we live with them, we interact with them constantly, they shape and confirm us, but we do not really ever perceive or understand them. Nor do we escape them. (Stuart Sherman is the miniaturist of this confrontation in his many "spectacles.") Language, history, science, are parallel. We impose systems (games) of coherence, of logic or even illogic (chance, reality-generating machines) on them, but they elude us. Life eludes us: we live it, but we don't know it. Foreman's spidery network of strings which envelop his stage and often the audience is symbolic of the fix we are in, as are his tricks with perspective, logic, and a bagful more, the syntax of his discourse.

In this production, feathers (something like the incredible lightness of being) dominate. But there are also crowns to suggest history; numbers, targets, and an oscillograph of Foreman's unnarrating voice to suggest science and technology (compare the failure of Ahab's massive dissection of the whale in his quest for essence); framing clusters of grapes

Originally published in *Studies in American Drama* 5 (1990): 91–93.

to suggest nature; primitive symbols to suggest myth (another kind of history); embroidery to suggest perhaps another dimension of human enactment (e.g., nostalgia, sentiment); and of course words, words, ironic words to suggest the impossibility of words, words Foreman, backwards, and inside out. All of it is presented with total self-consciousness, an awareness always of the derivative and deceiving nature of our thoughts, our impulses, our acts, our modes, and hence their invalidity (*Lava*'s male characters are all appropriately hunchbacks). (And why not the females?)

Lava, then, is Foreman's hot outpouring of what has obsessed him for many years, each play a different squirming. He *is* hysterical. He seeks to infect *us* with his hysteria by means of his surreal stage, his asymmetry, his sudden loud music and buzzers and bells, his subversion of our false contrivances of reality (plot, character, cause and effect, "grammar"). Here he offers a solution in his search for "the big guy" (or the big spider?) (not God, but what Melville sought behind the pasteboard mask of reality), namely to "try wearing a funny hat" (to "cool out"), Foreman's phrasing for getting out of the bind of our biology, our history, our language, our philosophy, etc. Of course it won't work. Foreman knows that. For one thing, it is too close to what he has already dismissed in illogic and chance. But a play has to end (or does it?). And the hat has feathers. Is hope the thing with feathers? For now, provisionally, yes. But not tomorrow.

Foreman, a postmodernist before that term's currency, is far too skeptical, feels far too deeply the awful mess of being human, to be really hopeful. Soon he will be shuttling again between the alternatives he poses in *Lava*: everything in this life is either a black hole or a mirror. Neither tells us anything true. But the struggle, the hermeneutic agonizing, *is* real for Foreman. We can be sure he will be squirming again soon, gaffed (like Beckett's Hamm) by his ontological dilemma. And we shall be gaffed by his staging of it. And perhaps between the two gaffings there will be a quick inkling of a raw feel (before the curtain comes down).

Michael Feingold

on *Eddie Goes to Poetry City*

"I think you're pushing me into very deep waters," the timid title character of Richard Foreman's *Eddie Goes to Poetry City* complains to one of the two young women who alternately lure and confront him throughout the piece. Her reply, snapped in his face like a rubber band, is, as you might expect, a brusque, "Swim!" When you head out on the ocean of Richard Foreman's thought, there's no escaping it: the boat of rationality is guaranteed to capsize, leaving you pleasantly adrift amid the shimmering waves of imagination. Foreman has been staging his everyone-into-the-mind-pool parties for so long that they've come to seem like seasonal festivals, twice- or thrice-yearly celebrations in which torments of the mind are made flesh, as a way of getting some brief surcease from the burden of consciousness. The tone and emphasis may vary from piece to piece, but not the substance; the metaphors shift their field, but the aesthetic strategies stay the same.

More affectionately comic than some earlier Foreman pieces, and with its characters more sharply defined as antinomies, *Eddie Goes to Poetry City* is no less nerve-wracked than its predecessors, no less startling to novice spectators. For experienced hands, it's an Easter-season treat with an exceptionally warm familial aura, and unusually spicy holiday cuisine. The recreation, as indicated, is as splashy and intellectually bracing as ever. In an interview I did with Foreman nearly twenty years ago, he spoke of his theater as the mental equivalent of a gym, where the startling lights and disjunctive actions were like exercises to enrich one's powers of imagination. In the new piece, his tone is more freewheeling, though hardly carefree; the spirit of the play is much

Originally published as "Farce Gratia Artis," in *Village Voice,* April 16, 1991, 105.

more evident, the effect much more like an afternoon in an amusement park than a dour, single-minded workout.

Foreman's playland, as always, starts off slightly forbidding in look: dark, heavy furniture (two long tables surrounded by chairs in the center, as for a board meeting); a floor littered with papers; ominous signs along the walls and unexplained objects hanging overhead. The ominous signs bear the legend "POETIC THEORY" and an anguished, openmouthed face; the mysterious objects overhead are Lucite letters spelling out the same phrase. This, then, is Poetry City, half workroom and half madhouse, with hints, at various times, of everything from a sex therapy clinic to heaven.

Where Foreman's scripts once went through a phase of prolonged defensive transactions with the public's resistance to his approach, now he invites them in: Eddie is Foreman-as-innocent, an ordinary guy who comes on proclaiming, "I want something out of the ordinary to happen," and to whom the omniscient Foreman voice overhead asserts, "You're less conventional than you believe." Eddie's conventionality, it transpires, is the field on which the imagination struggles to fly, sexuality rears its alternately ugly and tantalizing head, confusions of taste, logic, and purpose are sorted out or deepened, and, ultimately, as predicted by one of Eddie's tempters, a god manifests himself—bringing a celestial lunch, which is peeked at but never eaten. The Doctor, Eddie's spiritual counselor and / or diabolical nemesis, subsides into disempowered friendliness, and of the two women, Eddie seems to be preferring the passive-aggressive Marie to the openly challenging Estelle as the piece drifts to its balmy end.

To me the playful aspect of Foreman's work has always been its key element, helping it elude those who look for a rigorous philosophic schema, driving to distraction the squaresville types who demand a sequential naturalistic story. Here the playfulness is often quite literally both form and substance. "Office party!" the young women chirrup, dashing around putting up streamers. As befits the metaphysical talk and the descent of the god (a parodic Old Testament Jehovah in a long white beard), the first and third sections are light and cheerful. "You're big on speaking, but not on communicating," confrontational Estelle tells Eddie, to which his answer is, "That's because happiness floats into my language like a dream."

The floating dies down, but doesn't cease, in the more intense middle section, which focuses, appropriately, on Eddie's midsection, and his inhibitions regarding the intense matter of sex. The angst ratio is high, but the tone suggests Edwardian sex farce, with the hero bossed about by his women the way Rhoda, in earlier Foreman works, used to be shoved into French-postcard poses of the same vintage. "I never thought," says the demure Eddie, "I'd be showing such recent acquaintances my relative tumescence." The acting, lucid and funny in the best Foreman tradition, is at its peak here, with the encounter between Henry Stram's dazed Eddie and Rebecca Ellens's bristling, needle-sharp Estelle coming off like a Thurber cartoon, while the Doctor (Brian Delate) hovers over them with Mephistophelean delight and Marie (Kyle deCamp) sulks wistfully in the foreground. After such pleasures, rationality seems pallid indeed.

Ben Brantley

on *My Head Was a Sledgehammer*

Richard Foreman's theater tends to wither in the description of it. From the time of his earliest plays in the late 1960s, he has stocked his work with enough visual and verbal conundrums to keep disciples of abstraction babbling through years of exegesis, but the fact is that there is simply no satisfactory way to invoke what the experience of sitting through his plays is like. They exist fully only in the moment of their performance. And to try to boil them down to synopsis or allegory is to make them sound both pretentious and slightly bogus.

Accordingly, to watch *My Head Was a Sledgehammer,* the most recent offering of Mr. Foreman's Ontological-Hysteric Theater, is to be swept into a dazzlingly self-contained, thoroughly exhilarating universe that seems in the viewing—as does the best of Mr. Foreman's work—logical, rational, and disturbing in the way that individual dreams can be. It is a testament to Mr. Foreman's hypnotic artistic control that only afterward do you scratch your head and wonder what it was all about.

At its worst, this sort of abstract theater, even from Mr. Foreman, can leave one with the impression of watching someone else's hallucination and becoming annoyed and embarrassed by its self-indulgent opacity. In *Sledgehammer*—in spite of Mr. Foreman's continuing use of adapted Brechtian distancing techniques—one has the invigoratingly dangerous sense that the hallucination is occurring inside one's own head.

Sledgehammer, which has been designed and directed by Mr. Foreman on the tiny stage of the St. Mark's Theater, takes place in an environment that suggests a sort of nightmare classroom of the mind.

Originally published as "Metaphysical Lessons in the Dream Logic of Richard Foreman," in *New York Times,* January 19, 1994, C5.

103

There are three tall bookcases, their shelves stocked with objects ranging from disintegrating books to garbage-can lids and Jell-O molds; blackboards that appear to have been endlessly scrawled upon and endlessly erased, and rows and rows of white sheets of scribbled-on paper.

Staring at the stage is like looking at a drawing for children in which objects are hidden in an obscuring maze of lines. And among the play's joys is simply letting different props and aspects of the set drift, surprisingly, into consciousness, with the consequent impression that one has somehow called them into being. The process is compounded by Mr. Foreman's trademark use of strings that dissect the performing space, constantly forcing one to shift one's alignment of vision.

Moving through this environment are three principal characters, all portrayed with a masterly blend of otherworldly eeriness and visceral anxiety. There is the piratical-looking professor (Thomas Jay Ryan), a hunchback with a scratched face and an anguished, guttural purr of a voice; a male student (Henry Stram) who wears plus-fours and whose manner shifts between sycophancy and rebellion, and a female student (Jan Leslie Harding), who is dressed in layers of black and exudes a menacingly detached aura of sensuality. They all wear head mikes, and their voices seem to be emanating, disorientingly, from the same place.

Their dialogue has something to do with the forms of knowledge and perception and their ineffability. The professor, who is given to create rhymes "whose technique is they don't rhyme," announces at one point that he wants "to be the place through which truth passes." The admission seems both to torture and delight him. "Truth revealed," as he says subsequently, "takes the unfortunate shape of everything that isn't true."

In lines that often, in context, possess the shapely elegance of epigrams, the students bait and question the professor with unsettling mixtures of hostility and sensuality. "Do I have your permission to get torn to pieces by contradictions?" asks the male student, who from time to time seems to shift identities with the professor. "What does your internal time say now?" drawls the woman, whose cryptically sexual presence seems to arouse distinctly unacademic reactions in her mentor.

The pursuit of knowledge is a dangerous and frenzied process here. Papers and blackboard have a magnetic charge that violently draws and repels the academics. Abrasive bells and sharp flashes of light hurl the ensemble—which also includes four dunce-capped lackeys with Karl Marx beards who scramble on and off stage as a slapstick chorus of functionaries—onto the ground. Symbolic objects—as obvious as Edenic apples and messages on salvers delivered by the flunkies and as enigmatic as long baguettes, yellow golf balls, and single white gloves—appear and disappear.

Just as every statement in the play is answered by a contradiction, so does Mr. Foreman continually undermine our perceptual responses. The background music, which ranges from a soothing Philip Glass–like repetition of piano notes to ominously hyper, Felliniesque tunes, changes as soon as we become accustomed to it. And the lights dim and brighten as if the pupils of one's eyes are contracting and dilating.

None of this description can capture the ingeniously channeled energy nor the ecstatic humor that infuses the play, which takes the form of surreal visual jokes, precisely choreographed vaudeville routines, and, above all, the manifest frustration of its knowledge-seeking professor, who is, one presumes, a sort of self-mocking stand-in for Mr. Foreman himself.

Ultimately, there are no concrete answers in this endlessly mutating universe. Mr. Foreman, as always, seems far more interested in journeys than in destinations, in the intransitive rather than the transitive. And if *Sledgehammer* has a moral, it seems to be that to try to reduce life to a formula is to deny its confounding multiplicity.

It is important to note that the ways in which Mr. Foreman gives life to this metaphysical uncertainty this time around are purely, even brazenly, theatrical. Mr. Foreman may be constantly subverting the techniques he employs. ("This happens! This really happens!" proclaims one of the actors. "But if mere actors speak this, then it no longer happens.") Yet he has also brought them to a degree of extraordinary polish. After twenty-odd dogged years in the theater, he has clearly become one of its sophisticated celebrators.

Robert Gross on *I've Got the Shakes*

In the first minutes of *I've Got the Shakes,* Madeline X (Jan
Leslie Harding) decides that she must give a name to the theatrical
space in which she finds herself. She names it "ignorance," momentarily foregrounding, for the first of many times over the next hour, the
importance of unknowing in Richard Foreman's latest play. Madeline X
is the central figure in this theater of the unknowable, a teacher whose
self-professed lack of knowledge evokes both Socrates and, by extension, the Western philosophical tradition that has been built on the
shifting and ironizing foundations of Socratic inquiry. She serves as a
stand-in for Foreman himself, who transforms the workings of his consciousness into theatrical spectacle, without the aim of teaching any lesson. As Foreman explained to Elinor Fuchs, in an interview published
in the Spring/Summer 1994 issue of *Theater,* "As I get older I strongly
adhere to an almost anti-Brechtian stance in the sense that I don't want
to clarify anything, I don't want to give any ideas" (82). Foreman's
stance has produced this highly sophisticated parody of the Brechtian
Lehrstück, in which he leads his audience, not from ignorance to
knowledge, but to ignorance on a higher level of awareness.

Throughout *I've Got the Shakes,* Brechtian devices appear in parodic
forms. The famous Brechtian half-curtain is reduced to a long piece of
gray cloth which is held, waist-high, in front of the performers. The
Brechtian placard, reworked by Foreman into a strip of plastic that runs
across the length of the stage and on which the title of the play is repeated again and again, functions as a tautology, in which the play
refers back to itself. In Brechtian fashion, all scene changes take place in
full view of the audience. For example, fancifully dressed stage assis-

Originally published in *Theatre Journal* 47 (December 1995): 556, 558.

tants bring objects onstage, then hide them behind hand-held, red theatrical curtains, thus removing in advance any mystery as to what might be behind the curtains. And yet Foreman does not use this device to strip his production of theatrical aura, since the fact that the curtain stands in front of an object generates an anticipation for the moment when the curtain will be removed, and the object revealed. Thus, Foreman makes the act of unveiling, with its powerful theatrical associations, an event in itself. As another example, Madeline X's face is suddenly covered by a white disk, held by a stage assistant, and that disk, in turn, is eclipsed by yet another disk, held by another assistant. Since Madeline X is talking throughout this sequence, our anticipation builds toward the moment when her face will be seen again. Through such devices as these, Foreman gives strong aesthetic impact to theatrical gestures that momentarily obstruct vision and make the play reverberate with the power of concealment.

These devices find their verbal equivalent in the passages in which Madeline X talks about mystification. Mystification is rehabilitated by Foreman as a theatrical strategy that can help convey the complex texture of consciousness itself. Just as philosopher Nicholas Cusanus proceeded through a process of negation to contemplate the nature of divinity, Richard Foreman moves through negation in his consideration of human consciousness. As Madeline X meets with her students, Lola Mae Dupray (Mary McBride) and Sonya Vovovonich (Rebecca Moore), as well as with the hypodermic-wielding rabbi, Schlomo Leviticus (Michael Osano), she is confronted with a series of representations—a terrycloth rabbit, a shadowplay rabbit, pictures of a mountain, and pictures of a rose—all of which, upon examination, are revealed to be more absent than present. None of these are sufficient to give her understanding of her consciousness itself, and yet these inadequate objects are necessary objects of intentionality for her consciousness. Perception of the self is rapid, fleeting, and uncertain, as seen in the image of the rapidly flipping mirrors that the stage assistants manipulate—mirrors that only offer brief, fleeting, partial and intermittent reflections of the conscious subject.

Madeline X's philosophical dilemma finds its clearest expression in the photograph of the empty Ontological Theater that Madeline shows us. She is perplexed by the fact that she took the picture of the theater, and yet does not appear in the photo. She is pained and confused by her role as photographer—necessary for any representation to take place, but excluded from the representation, hidden behind the camera. For Foreman, a hidden, conscious subject is hidden outside the frame of any representation. "What is missing in this picture?" Madeline X's students ask about certain stage compositions. For those of us in the audience, we are missing from the picture, just as Madeline X is missing from her photograph. We might catch a fleeting image of ourselves in the flipping mirrors, but it would tell us nothing about the nature of our watching. The spectator is necessary to the theatrical experience but is eclipsed by the play itself, just as pure consciousness is always eclipsed by the objects of its intentionality. Madeline X's dilemma, we come to understand, is our own inability to know our knowing.

Foreman's emphasis on hiddenness throughout *I've Got the Shakes*—whether behind curtains, disks, mystifying language, or puzzling representations—turns the stage of the Ontological-Hysteric Theater into a site of Heideggerian *aletheia,* or unconcealment, in which the unconcealment itself still contains concealment. The lush profusion of objects that make up Foreman's set for *I've Got the Shakes*—Hebraic letters, playing cards, placards, strings of white beads, daisies, poinsettias, skulls, ceiling fans, clusters of baby dolls, a sketch of a woman's face, duplicated many times around the playing area, and the famous Foreman "strings"—are all easily identifiable, but their surrealistic juxtapositions always suggest that there is somehow more meaning to these objects than we can decipher. Our desire for coherence is aroused, and yet can never be satisfied, for the very ground of meaning is always hidden. By foregrounding the inevitable ignorance of the theater over its traditional claims to knowledge, *I've Got the Shakes* once again demonstrates Richard Foreman's role as the preeminent philosopher-poet of the contemporary theater. Deploying the devices of the Brechtian *Lehrstück* to move beyond it, he educates us in the unknowing that Bertolt Brecht sought to banish from the stage.

III Dialogues

Writing and Performance

Interview with Richard Kostelanetz

RK: What are your scripts like? And how do they differ from traditional theater texts?

RF: My texts these days are rather strange, and they are rather different from the kinds of texts I wrote when I began ten years ago. Ten years ago I would work from an outline, and the writing was the task of sort of deconstructing that outline, trying to attack the outline. If in the outline I had said, oh "Ben enters and complains that he doesn't like his job," I would, keeping that outline unit in mind, try to lie about it—try to write in response to that outline, lying about it.

RK: Lying about it? How?

RF: I would have Ben say, "I love my job." It was a matter of a kind of spiritual discipline, keeping in the mind what I knew was supposed to be the content of the scene according to the outline, and then perversely trying to put on the page all the negative, the reverse sides. . . .

RK: To change the detail or to invert the detail with respect to your original scheme.

RF: Yes. Now writing plays in that way, I found that every couple of months I would write a play. Before that period in which the writing went well, day after day I would sit down, start to write the play, and there would be a false start of one sort or another, and I would throw it

Originally published in *New York Arts Journal* 28 (December 1982): 16–23 (excerpt).

out. At a certain point—I suppose in the middle 1970s—I decided, "Why should I throw out all this material?" These false starts express the real energies of my life, the real frustrations and successes of my life, and I began staging this day-by-day effort to write.

RK: Your *own* effort to write?

RF: Yes. And philosophically at this point I cannot conceive, for the theater, of staging, of writing anything other than these kind of day-to-day notions of my struggle to write. It seems to me that building a more coherent, consistent structure, a more unified structure, is a lie. It's a lie about what my life here in New York City in this moment of the twentieth century is about. Now when you look at my texts these days, not only are there page-by-page radical changes of locale, of idea, of what-have-you, but there are no characters indicated. I am not writing lines. I am writing information. There is no indication of the context in which this information occurs.

RK: Now doesn't this relate to what Gertrude Stein was doing in her theatrical texts, which likewise by and large eschew character, setting, and all specific connection between particular lines and certain characters?

RF: Yes. My work began specifically under the influence of a French philosopher, Gaston Bachelard, and I began with the notion, suggested by Bachelard, that my writing would be the registration of a certain level of activity. In my case, I decided in my first plays only to talk about what was physically experienced by the body, to wipe the scripts clean of any ideas, any conclusions; only sensory impact went into my scripts. About two years later I started reading Gertrude Stein for the first time and became very influenced by the notion of writing exclusively in the present and in response to present stimuli when you were writing. I remember Stein's original statement in one of her lectures that the thing that always bothered her about the theater was that what you were watching was always ahead of or behind the time of what was going on stage.

RK: What does that mean?

RF: Well, it means that watching a play is not like reading a book, where you control your own time, control the rhythm of your own aesthetic experience. It means that you are always leaping ahead anticipating what the character is going to do next and character B, responding to the statement of character A, is either saying after you imagined the response or before you imagined the response, so it's a syncopation in time.

RK: Isn't that one of the things that makes theater interesting?

RF: Yes, well, I was about to say that Stein, of course, created what she called kind of "landscape" theater because this bothered her. I tended to do the same thing for a while, but in the last few years, if I've done anything, it has been to exploit all the more radically, I think, the nervousness of not being together as spectator and stage life. And I am exploiting the fact that I, as a writer, when I am writing, am always ahead or behind my own writing. And I am allowing the normal stumbling, the normal difficulty, the normal incoherence that results from that split time to interpenetrate my text and to disrupt my text.

RK: But then your writing is in part about the perceptual experience of writing—about the perceptual experience of understanding and manipulating language.

RF: My writing is about taking dictation as enunciated by the American poet, Jack Spicer—just taking dictation. At a certain point lines start to come from somewhere; but not letting it come through purely, allowing the failure of my own mechanism as this radio that is receiving dictation to interfere with the text. There is a collision between the spirits, the muse, and the fellow that's sitting here with a heavy pen, with a tired hand, with a tired brain.

. . . . .

RK: How do you realize your texts?

RF: These days there's a text that is simply a series of lines with no character indicated. The first thing I do is to design a set. It's difficult to say how I decide to design a set, how I decide what form the set will take. When the time comes to produce a play, I have many, many texts that I've *generated* (I prefer to say rather than written) over the past few years. I very casually pick one. I pick them on the basis of a very casual reading of the text, you know, just skimming through it. Then to design the set, I might look at it over the period of a minute and a half, just getting the feel of what some of the scenes are like, some of the pages are like—and then I start to make sketches and just sort of doodle my notion of articulating a kind of space. As the set proceeds, it sort of takes on a life of its own in a certain sense, in a way similar to the way that Cage and Cunningham, the American composer and dancer, work when they're creating dances, where Cage will perform a musical score that Cunningham doesn't listen to while he is creating the choreography.

RK: Except that you are both the author and the stage designer here. Are you trying to separate your functions so that they both function autonomously?

RF: What I'm saying is that when I'm designing the set, I don't go back and refer to the play very often to make sure that I'm doing the right thing. The set gets built. Then rehearsals start. And when rehearsals start, my task is to find a way to make my text exist in the world of this set in much the same way that the various animal species on this planet have to find the way through natural selection to adapt to the circumstances of this planet.

RK: But then you're competing against yourself literally, or one part of you is competing against another part of yourself, or one expression of yourself. . . .

RF: We're not competing. We're falling over each other. Because I believe that in mistakes, in falling-downs, in these collisions, unexpected collisions, there alone arises the opportunity for invention, arises the opportunity for creation. So I try to create situations where things will collide and fall over each other, and get into trouble in that way.

Now I suspect that most people think of me as a director and say, "Oh, well, you have an ability to manipulate space, to be an effective theatrical director, and your texts, of course, are sort of nonsense in a neo-Dadaist tradition." I feel that that is absolute nonsense and an absolute misunderstanding. I think of myself as a writer. I feel that in all kinds of ways my writing is much more adventurous than my staging, which, I fear, tends to take these rather exploratory texts and turn them into a more classical kind of theater than I'm totally satisfied with.

RK: In what sense do you mean the writing is more exploratory? It's less conventional, less immediately accessible?

RF: I would like to discover, to build for myself, a way of staging plays that exploits this same kind of disassociated, daily note-taking, that exploits and uses these false starts, that is very honest about what comes up in my fallible human life. I would like to stage in a similar way. I find it very difficult to do. I find myself invariably making it work better in classical theatrical terms.

RK: Would it help to have a text of your play while seeing it? Would that give the language more emphasis?

RF: Lord, I have no idea. I can't possibly say because I'm not thinking very much of my audience when I am making this stage a theatrical object. I'm trying to create a device, a machine, that will have a certain effect upon me when I see it, and then it is up to other people to use this machine or not to use this machine as they see fit. I must say that in staging my work, I do all kinds of things to make it more difficult for the audience to perceive the text. For instance, let's say that during rehearsals we have determined that Rhoda, who is my leading character in all of my plays for the last ten years or so, says, "I'm going to the store to buy some whole wheat bread." I will record on tape four different voices, one of them Rhoda, three of them other actors. Each of them will have one word from that sentence—I'm . . . going . . . down . . . the . . . street—each of those words by a different actor. Each of those four actors has their own loudspeaker.

RK: Since you have four-track tape. . . .

RF: There's a loudspeaker in each of the four corners of the audience, so you hear this line word by word, the first word coming from lower right, the second word from lower left, and so on. At the same time this is coming over the loudspeaker, on stage the real live Rhoda will be repeating that same line at a different rate of speed from the words coming from the loudspeakers. Now, why? Not simply for the music, though partially for the music of such a sound. I'm interested in the theater, in freeing individual elements, freeing the individual word, freeing individual gestures, freeing individual noises in much the same way modernist poetry since Rimbaud tends to free the word on the page. So that in the third sentence there is a word that is a little loosened from its moorings, so you can relate that word to a word back at the beginning of the poem and not read it in a direct narrative line as it comes up. Now it does make it difficult, therefore, for the audience to keep on top of what exactly is being said. I have profound faith that on some unconscious or semi-conscious level an awful lot comes through, even though we may not really be aware of it.

RK: Because of the strength of the language or the strength of the articulation with the tape machine and the live performance?

RF: Because of the strength of the articulation of the entire space-object, which is the performance.

RK: Could somebody else realize your text?

RF: Oh, yes, as a matter of fact, I think there have been five different productions of my works that I have not directed. I'm not too happy with them.

RK: How were they different from what you would do?

RF: Well, I thought they weren't different enough. I would hope that some day someone would take my texts and do a production radically different from anything I could have imagined for them. These productions were too close to what I could imagine for them.

. . . . .

RK: Why theater?

RF: Oh, I've always had very ambivalent feelings toward the theater. I started making theater when I was very young because I was a very shy kid and it enabled me to live a less shy, more active life. So when I was ten years old I started making my own plays. As I grew up, I grew more and more unhappy with the medium that I found myself placed in. I harbor tremendous distrust for the theater in the sense that the theater classically and almost unavoidably is an art form that is made for a group of people together to respond to. There is one thing on this planet that I don't trust: it's the response of any kind of group of people. If I could take the individual members of my audience and work with them, I'm sure I would find them to be more intelligent and more sensitive than they're going to be as a group.

I do love the kind of three-dimensional manipulation in space of all the elements. I theorize a great deal about the writing of texts and the implication of texts; however, putting actors on a stage and manipulating them in three dimensions with scenery I have no theories about, because it's something I just love to do and I do it very easily.

RK: You've collected your texts in a book, *Plays and Manifestos.* How should it be read? Is it just documentation or is it indeed scenarios that can be performed?

RF: Well, there are scenarios that can be performed. I believe it's literature also. I'm primarily concerned, I would have to reiterate, not with the theater, or not with film. If I make film, I'm concerned with a kind of activity that has something to do with my experience when I am thinking and get ideas. When I get ideas, that little gap is jumped by my head and by my whole physiology. Something happens in my body, and I am concerned with trying to make an object—make a machine that evokes, is a metaphor for—that somehow operates in a similar way.

RK: And so a book of your texts can be this kind of machine?

RF: Yes.

Both Halves

of Richard Foreman: The Playwright

Interview with David Savran

DS: Who are the artists who have had an influence on you as a writer?

RF: Nobody was an influence for a long period of time, except Brecht. For about twenty years Brecht was, I suppose, my god, and I tried to find out everything I could about him. His rather flat-footed, aggressive literary style influenced the way I wanted to write. I thought his way was obviously *the* way to work in the theater. He was absolutely my greatest influence. He certainly is not any more.

DS: What was the effect of your studies at Yale?

RF: I went to Yale as a playwright and wrote, one year, my imitation Arthur Miller; one year, my imitation Murray Schisgal (believe it or not); one year, my imitation Brecht; one year, my imitation Sartre. I studied under John Gassner, whom I found to be an extremely rigorous teacher, although even at that point my aesthetic was somewhat different from his. He thought I had talent, which was gratifying to me, but he said my one problem was that I tended to go for a big dramatic effect and would then repeat it and repeat it and repeat it. For a long time I tried to correct that until I realized, "How silly. If that's what I want, I should turn it into an asset, or at least try to." And indeed, that describes, sort of, the structure of the work I've been doing since then.

When I left Yale and came to New York, I met the underground cinema people. They were interested in contemporary American literature, which, up until then, I had denigrated completely. I thought of myself

Originally published in *American Theatre*, August 1987, 14, 19–21, 49–50 (excerpt).

as being a rigorous, vicious European intellectual and had contempt for what I thought was the more primitive, naive American approach. I was introduced to Charles Olson, Robert Duncan, the whole school of poetry and, through them, to Gertrude Stein, who, when I discovered her in 1964, became the second big influence in my life. It was really the combination of Stein and Brecht, I suppose, that produced the first thrust of my work—with one other bizarre contribution: Because of the interest of some of these filmmakers in various esoteric matters, I got interested in alchemy.

I began thinking of the attempt to write a play as the attempt to work and rework the same material, much as the alchemists would keep working on their combined metals to transform them into gold. I really thought of writing a play as taking certain basic physical givens of the situation in the play, repeating them with slight variations again and again and again in the text.

DS: Speaking of repetition, formally your work proceeds not through a linear, causal development but through a series of variations.

RF: Yes, that's the way I think. Of course, these earlier plays—from about 1968 until about 1975—were a bit different from what I'm doing these days. Now the writing tends to break down into very aphoristic fragments of, I think, a kind of elegant language, even though there's no narrative progression. When I began there was no witty, aphoristic style. The characters spoke a very flat-footed notation of the physical sensations in their bodies: "Oh, my hand is now heavy. It's still heavy. Why is it heavy? Why does it feel like it's growing?" Discussions like that would go on and on and on, and circulate. And then there was a shift. It was almost as if I wanted to start from scratch, to start from the ground of matter before I dared to use a more aphoristic, comic style, which spoke more openly about all of my real metaphysical concerns.

DS: Were you interested in using Brecht to provide that concrete, material base?

RF: I wasn't thinking of that aspect of Brecht, I was thinking of him totally in terms of the alienation effect, the desire for nonempathic theater. The language, meanwhile, was derived from this alchemical approach to manipulations. Stein also, in trying to write in a continual present. So the move into a more aphoristic style was a move away from that continual present into recycling all the inherited garbage and treasures of Western culture, which I think has been the emphasis of my work in the last ten years or so.

DS: How do you start working on a piece?

RF: Well that has changed radically over the years. First of all, I work as a poet works. I think that's the cause of one of the problems I've had with the critics. I don't think people who go to the theater spend much time pondering contemporary poetry. If they think of another literary form, they think of novels. They do not think about the implications of language and what can be done with it. And language, rather than narrative, has always been my concern, along with the psychological implications of what the words do to the person who is speaking them, the way that they hit associations and strike off other trains of thought.

John Gassner taught me that plays aren't written but rewritten. Actually, it used to be a lot of fun. I would fix and fix and fix and fix. When I began writing the style of plays that I write now, I still believed that a play had to be written from an outline and then rewritten. But I came increasingly under the influence of theoretical work by artists about how twentieth-century art should be produced. I decided that it was perhaps interfering with other sources of creativity to work from the constriction of an outline.

So I began writing without an outline. However, what happened was that I would get an idea for a play and start to write it out in scene form and after two pages, ten pages, sometimes twenty pages, it would dry up. About once a year—after sitting down every day to write—suddenly a play wouldn't stop and I'd go through to the end. That left me with notebooks full of plays that never went anywhere. I remember saying, "Why can't I use all theses false starts and actually stage them?" I believe the first play that was made up of false starts was *Hotel China*. Eventually I began to feel free to take pages from last year's work, shuffle them around, treat them as raw material to manipulate. That, essentially, is the way I work now. For many years I was very shy about admitting how loose my compositional methods were getting. I began to feel guilty, thinking, "You're not sitting down and working like a good solid, honest workman, the way Ibsen did." So I didn't come right out and admit my increasing belief that the most interesting writing would simply come to you at odd moments. You had to pick it up on the fly, as it were, and then treat that random material much as you would treat the random material of your life, selecting and organizing all kinds of impulses in order to write an Ibsenite play.

This is not at all different from the way that most twentieth-century poets work, though I suspect not too many novelists do. The reason I work this way is simply that I have different metaphysical interests, different therapeutic interests in making art, than Herman Wouk might, if he's writing a novel that he hopes is going to sell a lot of copies and be bought as a movie. In the last year and a half, I have tried more and more to integrate into plays song forms that I've been writing to specific tape loops. I write under the influence of that looped music very repetitively, and try to let come whatever comes through the motor input of those musical rhythms.

DS: When you're writing a play, how do you conceive characters?

RF: You have to understand that I don't write a play anymore. I put things down in notebooks. And when it comes time to produce a play, I look through the notebooks to find interesting language, interesting sections. I haven't written characters except perhaps during the first years that I was writing. These days, when I give a text to the actors there's no indication of who is speaking. It looks like a poem on the page. For the first eight years or so, even when I wrote in the notebooks, I would write down who was speaking. I felt sort of casual about it because I had always believed along with Max Jacob, the French poet, that character is an error, that our characters are determined by the accidents of our birth and our social circumstances. If you go a step further, you can say that they are the accidents of our genes. I have always been interested in trying to write from and evoke that level of the self that underlies character, that level of consciousness that we all share, upon which is superimposed the accident of character.

When I assign the lines, however, I think of the collision between the particular character of my performer and what I consider to be the more universal thrust of my language. So I will say, "How does this universal statement about fear collide with the particular characteristics of Kate Manheim, or of [the late Wooster Group actor] Ron Vawter, who has other characteristics?" Let's say that Kate and Ron were in a play together. I could totally reassign the lines, I could restage the play and I profoundly believe that it would be just as true, that in staging it

I would find other, just as rigorous solutions with that distribution. For the last ten years, the scenario for my plays has evolved in rehearsal. Not the language—that stays pretty much the same.

. . . . .

DS: What impact does your work as a director have on your scripts?

RF: I imagine that every artist dreams of creating in the most organic way possible. I don't know if that means that ideally I would like to write the play in rehearsal. But I do know I specifically leave out many narrative keys in my writing because I feel they can be more suggestively, imaginatively fulfilled in the staging. I know I'm staging it and it's part of the same process. Whereas when I pick up Ibsen or other, more contemporary authors, I often have the feeling, "It's all here in the text. Why bother doing it?" I think that's one of the reasons a lot of directors think, "Well, I'm going to do *Hamlet* but it's going to be done under water." *[Laughs]* How boring just to do it unless you can do it differently!

There are a lot of people who have said "I like his directing—well I don't *like* his directing, his directing's okay—but his texts, well, there's not much to his texts." I have confidence, though, that in time people are going to be able to relate to my texts and see that they have the same coherence and density as a lot of twentieth-century poetry. I'm not saying I'm as good as Ezra Pound. I am saying that my texts operate in the same way and are not incoherent and meaningless. The fact that I generate them differently from the way people usually generate texts for the theater is, I think, quite irrelevant.

I have often said that I'm a more conventional director than I am a writer. I think my writing pushes further ahead, faster, in terms of stylistic, aesthetic adventure, than does my staging. I think I tend to domesticate texts that are wilder than my staging.

DS: Do you read the critics?

RF: Yes. But I've always wished that I didn't. I hope that maybe this year I won't. I get very upset, as I think most people in the theater do, when they are negative, even though I may think the critic's a fool and what he's saying is clearly foolish. I agree with Gertrude Stein that artists don't need criticism, only encouragement. All you can do as an artist— for the kind of artist that I am and that I think artists should be—is to try to radicalize your own impulses and strip away everything that isn't you and make whatever *is* you that much stronger. Of course, critics who write for the newspapers are interested in talking about how well the audience has been manipulated, not about how rigorously a specific vision has been pared away and presented onstage. So it's frustrating, but that's life.

Now that we're moving into a more conservative political era, many have criticized me and the kind of director and writer I represent as being self-indulgent. I'm accused again and again of solipsistic vision. I find that absurd. It simply means that the people reviewing me don't recognize my cultural references. Moment by moment my plays are appealing to this or that cultural reference. For instance, *Africanus Instructus* is about the discovery of Africa and about how the modern world destroys all the exoticism of foreign cultures, about how the energy goes out in this collision between societies. There's one scene in which a black man from Africa brings in a red telephone on a platter and the actors start singing about this red telephone that means emergency to the white people. A critic for the *Village Voice* wrote, "He just picks out of his unconscious these meaningless images and they're singing about a red telephone. What does that have to do with Africa?" As if a red telephone doesn't have to do with all the hangups of the West and the hot line to Moscow and our emergency-oriented life! All of the symbols have reference to culturally inherited preoccupations. Maybe the critics just aren't as well read or as well informed as they should be.

DS: I also think it's because they're not used to dealing with such loaded images outside of a narrative framework.

RF: Yes, I suppose that's true. But if you've had any exposure to the other contemporary arts, you know that is the way material is organized these days by artists.

DS: How aware are you of the social implications of your work?

RF: Immediately—after the fact. Maybe there are some artists who proceed purely by calculation. I don't. I talk about my work afterwards, and I can analyze intellectually as well as anyone else what is operating in my work. But I think an artist sort of does it, and then sees what he has—and he either throws it out or goes on. That's the way I think all artists work.

DS: What are your goals for the future?

RF: I have absolutely no goals anymore, except maybe to get out of the theater. I think one should change, and the remaining opportunities for change in the theater are not that great.

I've always thought that my work (and most interesting twentieth-century art) is therapy on some level. None of us live anywhere near the potential that is built into us. We're all asleep, me included. If you're a good artist, your art tends to be better, more rigorous, more alert than you are when you're alone in your apartment eating dinner. If I have any goal, it would be to make some of the rigor of my art pertain more in the other hours of my waking life. I'm very aware of the fact that when I make a work of art, I'm making a place where I would like to be. I wish the world had the same rigor that my work of arts has, and I don't think that's a particularly healthy wish. I realize increasingly that to hope that a work of art is going to solve your life's problems, or any life problems, is not very sensible.

I now deeply, fully understand that no matter how good my art gets, it's not going to change anything. And that realization, sinking into me slowly, has had a profound effect on the way I think of the future.

DS: Do you hold out any hope for the theater?

RF: I wish more people who are concerned deeply with language would try to work in the theater. I think, for instance, in America now there's a young group of poets who are very great artists, part of this so-called language movement in poetry.

I think I will always be interested basically in writing—in generating language that can be mostly talk and then some gesture, music, staging. Or the balance can change. But to me, it's all the writing of a text, whether or not the staging is a part of that text. So even as a director, as a person who makes his scenery, makes his music, I think of myself as a writer making texts.

Both Halves

of Richard Foreman: The Director

Interview with Arthur Bartow

AB: At first, the theater thought of you primarily as a playwright. Now there is the sense that the director side of your ego has taken control. What separates these facets of your talent?

RF: I am driven by a desire to put something into the world that I find lacking in my life, and I try to correct for that lack by making works of art that give me the environment that I would rather be living in. Originally, I became a director simply because nobody else would direct my plays. I didn't really direct any material other than my own until I did *Threepenny Opera* at the New York Shakespeare Festival in 1977.

AB: How does a young man from Scarsdale who goes to Brown and Yale in the 1950s, and then goes to New York to write Broadway comedies, become a unique revolutionary artist in the theater?

RF: For some reason, from the time I was very young, I had an attraction for the strangest material. I read *The Skin of Our Teeth* for the first time when I was about twelve and thought, "It's like a dream, it's so weird, it's wonderful." I remember seeing Elia Kazan's production of *Camino Real,* which I dragged my Scarsdale parents to, and they said, "What's this all about?"

I have always gravitated to things that try to talk on some more spiritually oriented level, rather than realistic discussions and manipulations of the real, practical, empirical world in which we live.

The big revelation was discovering Brecht—and especially his saying that you could have a theater that was not based on empathy. For some

Originally published in *American Theatre,* August 1987, 15–18, 49 (excerpt).

reason, even at an early age, what I hated in the theater was a kind of asking for love that I saw manifested on the stage, getting a unified reaction from everybody in the audience. Brecht said it didn't have to be like that, and until I was in my mid-twenties, he was the beginning, middle and end of everything for me. That only changed when I came to New York and encountered the beginnings of the underground film movement. I discovered that people my age were making their own movies, operating on a level that was akin to poetry rather than storytelling. And I thought, "Aha! Why couldn't the techniques of poetry that operate in film operate in theater?"

AB: Your plays have been examined and explained and analyzed in an effort to understand their meanings. As a director, do you feel the need to insert guidelines in your work to give it clarity?

RF: Yes, more and more. Oftentimes, however, people find those explanations confusing. First of all, things change and we get older, and I don't know if what I do is quite as hard to understand these days as it was fifteen years ago—because other people are doing similar things. Other currents of thought are in the air.

In fact, some of my explanations are couched in terms that are rather difficult because to be true to what I'm trying to get at *is* difficult. I don't think of myself as doing anything radically new or different. I think of myself as being a meeting point for all kinds of ideas, all kinds of feelings, that are around us at hand. It seems to me that I'm dealing all the time with things that are in the air, both as a director and as a writer. I take them and I try to play with them in an exhilarating way. Part of the difficulty is that people are sitting there thinking, "Yes, but is he saying that we should be this, that, or the other thing?" I don't think that's my function as an artist. My function is to enjoy and help my audience enjoy an exhilarating kind of play with all of the elements that are present in this very heterogeneous culture where we have hundreds of years of history—and everything that's been thought and felt in all those hundreds of years—readily available on bookshelves, TV, cassettes, records. It's all there as never before, and how do you keep your head above water? Well, you learn how to ice skate on the crystallized surface of the pond, underneath which is all of the morass of these hundreds and thousands of years of history.

You're trying to be a medium, to let all of these messages come through. Your task is to make some kind of harmony out of them. You do have to eliminate some of the noise so that something is perceptible. I have never been in favor of the kind of big undisciplined soup theater that for me a lot of the 1960s mixed-media things were. Even though many people can't perceive it, I've always been an extremely structurally oriented artist who tries to clean everything up. As a director, my one criticism of myself is that I'm too neat—I've tried to keep the event too clean, too defined, too much under control, and I wish at times that I could be a little messier as a director, as I am in my writing.

AB: Is this cleanliness, this structure, tied to the way you work with actors? You've said before that you tend to be dictatorial in terms of detail, movement, placement, and timing. All of those details seem to come exclusively from you.

RF: This is less true today than five years ago. Unless I am totally deluded, these days the vast majority of actors I work with find it a positive experience, whereas in the early days some of them definitely felt very constrained. I still block the play and I still ask for all kinds of specific things. I never ask an actor to do something he's uncomfortable with.

AB: Your use of space and the scenography for your plays is also "neat and defined," to use your words.

RF: There's no denying that my main interest in the theater is compositional, that I am interested in the interplay of all the elements. I am not interested in the theater where the audience becomes seduced by a kind of empathetic relationship to the actors.

AB: And yet, in a sense, audiences are seduced because there's so much humor in your plays. We laugh, we find it amusing.

RF: I think people should laugh more. Audiences are often afraid of laughing because (a) things are going so fast that they think they might miss something, and (b) I come with this reputation of being tough and intellectual and radical. I've had friends of mine sit in the theater at my plays laughing, and people sitting in front of them look back at them as if it were some sacrilege.

AB: Your famous wires stretched across the stage seem to suggest an alchemy, a science. It has come to the point where a designer can't put a wire across the stage with having it referred to as "a Richard Foreman wire."

RF: Yes. But all those things of mine are slowly disappearing, at least becoming more minimal. When I did a revival of Arthur Kopit's *End of the World with Symposium to Follow* for American Repertory Theatre, we were starting to rehearse, and [ART artistic director] Bob Brustein came and sat next to me. He said, "You know, this is really going well. But tell me something, are you really committed to those strings?"

AB: With those taut strings, are you trying to frame stage areas, to bring them into more concise focus? Visually, there's somehow a connection with those drawings in antiquity where human figures are extended along straight lines.

RF: It's hard for me to talk about, really, but somehow there is an articulation of the space—so that it's almost as if the actors are overlaid by a kind of grid. It's almost like a musical staff. I just like knowing where I am physically and, somehow, also on a spiritual and emotional level. Those strings define some kind of force field, some kind of reverberation box within which whatever is going on in the play reverberates even more intensely. It's like someone sketching who may start making lines of force, feeling the need to work within a diagonal, and then the body grows out of that.

AB: Almost a connection with God.

RF: Well, it does connect with God in that it connects with what I think are the most abstract spiritual energies—the kind of nervous motor energy that wants to find a way to concretely manifest itself in the three-dimensional world. It starts out as impulse. And my technique in the theater is to feel the impulse, not knowing yet what it means or how it wants to work, but to let the impulse lead me. Then it takes on a three-dimensional, actorly, prop-like form—but I remember to always keep present for the spectator a kind of interplay between the original thrust, the place that it came from, and the real three-dimensional human, physical manifestation that it takes on at this particular historical moment. The impulse leads to another impulse, which leads to another impulse. I work the same way when I'm doing Brecht or Molière or Kopit.

AB: That's easier to envision in your own plays than in something like *Don Juan* or *Threepenny Opera*. In those cases, how does that impulse manifest itself in you as a director?

RF: When you're actually working on something, you don't think conceptually. At least, I don't. It's important for me to delineate the way in which the play grows. The first thing I do is make a set. These days I generally design the set even when I'm working with a designer—I basically build a model that I give the designer. The set for me means creating a kind of space that both implies the grid of this original thrust, the energy of this original abstract thrust, as well as the specific locale of the play. Given the proper set, I then have within it different layers of being so that this impulse can realize itself.

In addition to strings, a lot of my sets use railings and specific geometric divisions of space that might suggest a courthouse, a bullring, a synagogue. Now all of these enclosures somehow immediately give me a grid-like place within which to work. That's both the suggestion of the impulse of wanting to present things to your fellow man plus defining special sacred private places where you are alone with your soul or whatever. At the very root, it allows me to play with more public impulses as well as more private impulses within any text. That's an oversimplification, of course, but it seems to me those are the energies that are set up.

The next thing I do is get music organized. All my plays, including the classical plays and the contemporary plays that I haven't written, have music behind the text most of the time. Those are taped loops that I make myself, that I take from a variety of sources. Again, I choose music that somehow takes a section of the text and makes a comment on it or lifts it into a slightly different plane.

AB: Do you sometimes choose music that is in opposition to the action of the play?

RF: Oh sure. That's an old Brechtian technique, of course, to distance, to estrange. But I don't think about that conceptually. It's a way of working that happens almost automatically. At the moment, I don't tell myself why it's interesting. It just seems right to me. We start rehearsing from the first day with all that music.

AB: Your major collaboration has been with composer Stanley Silverman, working on some five operas, including *Dream Tantras for Western Massachusetts, Hotel for Criminals,* and *Dr. Selavy's Magic Theatre,* all of which you directed.

RF: The productive thing about Stanley and me is that he understands completely what I'm doing. In that sense, he's one of the most perceptive and intelligent people I've ever worked with in the theater. We have somewhat different tastes, and I think he isn't as interested in some of the really far-out things I'm interested in. We generally have one or two meetings where I give him the words and he's interested to find out what I think is going on and what kind of music I imagine. That doesn't necessarily mean he would write it that way. We're friendly, but we've never talked that seriously about anything.

AB: Then the setting, the music, and all of this has been digested in advance of rehearsals. Is that basically a subconscious gestation?

RF: Absolutely. Before I go into rehearsal I know my stuff, but I don't sit pondering over the play. I read it once or twice and make very brief notes about staging ideas in terms of the set I've designed. But they're all tentative. Any text includes hundreds of possibilities. Until I hear the specific actors that I am using, I don't know in which direction I'm going to be logically led. I will discover certain things they are emphasizing or that seem to be true because of their personality that relates to certain possibilities in the play. Then my task is to strengthen that line of interpretation.

Of course, I do have a clear idea of what I think the play is saying and the direction I want to take it. It's like being shipwrecked on a new planet. How do you live on this planet? It demands certain things, there are certain rules you have to abide by to live on this planet. But, within that, you could build a house with five rooms, a two-story house or a lean-to. Those are the decisions you make in the rehearsal period. The only way I know how to work is three-dimensionally, to get up immediately on your feet with the actors and feel things in the body and make it happen that way. It's my articulation of the actors and space, vis-à-vis the text psychologically, that I'm proudest of, and I have absolute total confidence that I know how to do it.

. . . . .

AB: You once said, "Style attacks with truth—where man most deeply is, but where he has the least developed navigational techniques."

RF: Believe it or not, even when I was fifteen years old and used to go to Broadway theater every weekend, I hated all the hits but occasionally I'd see flops that I thought were wonderful. Americans just don't understand that style can be content, and style has things to say. Invariably, people never understand that in art, stylistic position is a moral position, an intellectual position, and carries the real content, the real meaning. That lack of understanding is continually frustrating. But it's easy to see why it's so difficult. In order to live your life in a normal capitalistic society, you have to put blinders on so that you are not distracted from the things that you have to do to get on in the world.

AB: To add to that conflict, your early productions shocked audiences with sound and lights.

RF: I think people still tend to find my work abrasive and aggressive. I still use lights in the audience's eyes. I don't use loud buzzers anymore. But a lot of that stuff is still there because I want to wake the audience up, to stop them from being seduced by what they're watching. Lucidity, clarity, waking up. That's what I'm interested in, both in my life and in my art for myself and my audience. I've discussed this with some other modern directors like Elizabeth LeCompte [of the Wooster Group]. We make these things up because it makes us feel better. It's not to torture people. It's to feel good, like after you've had a workout in the gym. But, of course, by doing that you sometimes run into hostility.

I've been very lucky because I've been able to do exactly what I wanted to do in the theater for twenty years. That quest is an attempt to bring onto stage the operations of some other energy that is not the energy of the human—the socialized human personality. I am trying to do it, believe it or not, through rhythm. A lot of people have always said, "Foreman is basically a visual artist." I am interested in a kind of dialectical relationship between what you see and what you hear, which becomes a kind of rhythmic articulation—an evocation of a different level of being, a different kind of energy that one can bring into life.

AB: Why have you extended yourself to other work besides your own plays?

RF: The reason I'm doing plays other than my own now is to see whether this particular kind of rhythmic articulation is applicable to all kinds of works. One way to relate it is the old theory of the Jewish cabala, that the world we live in is a world of broken pieces of physical material, reality, and our task as human beings is to somehow find the spark of light in these things and lift them back to God, to the wholeness that they're supposed to have. I know how pretentious that sounds, but I'm trying to take things that show the picture of our fallen physical world and find a way to organize it rhythmically so that somehow it starts to swirl and starts to lift and some other quality comes through that restores it to something else in the cosmos.

AB: Do you find yourself limited by your own frame of reference when directing your own work?

RF: When I am directing my own material, I have no inhibitions about treating "Richard the author" as a joke, with contempt, making fun of him all the time, making fun of my text. I automatically do that with all texts that I'm working on. So I have to watch my Ps and Qs if an Arthur Kopit is around. It doesn't mean that I don't respect his text, but to me you've got to play with this stuff, you've got to handle it just like it's just stuff. It's not holy. The great relaxation of dealing with one's own scripts is you can say, "What is this garbage this guy wrote? How are we going to fix this mess?"

AB: You've worked abroad a great deal in the past few years, but now you're back in this country for an extended period. Is that because you feel that the pendulum will swing back in the next few years to where audiences will be willing to hear all kinds of disturbing, less rational truths?

RF: I think I'm interested in America because I had a problem most of my life in wanting to cast out all of those traits that I didn't like about myself. And one of the traits I didn't like about myself was being, to the bottom of my soles, an American. I feel that the American culture is an adolescent culture. I feel that I'm an adolescent and I idolized what I thought was the greater maturity and sophistication of Europeans. I wanted to identify with that but, finally, it ain't me. I have to come back and work out of the dumb, naive openness that is a great strength of America, but was very hard for me to accept.

Off-Broadway's Most Inventive

Directors Talk about Their Art

Discussion with Elizabeth LeCompte

RF: I once read an interview with Ron Vawter in which he said that you could make a play out of whatever happened to be in the room in which you found yourself—whatever was there at hand, you'd compose a play.

EL: He said that about me?

RF: Yes. That you work specifically with what's assembled around you. Isn't that how many contemporary artists work? I know I do—even things you think are coming from someplace inside you—it's still material that your circumstances have provided you with. But then there's the next step. . . .

EL: . . . the choice. How some things get chosen. Let me ask you—did you ever think, "Alright, I'm going to have someone take his clothes off and show his asshole? I'll just feed that to the lions? And then I'll go on, sort of as so many artists have done?" Do you think of it?

RF: Yes. Because I work very differently than you. Even though I think I'm very gentle and I've never asked someone to do something they didn't want to do, mostly I'm the one doing the improvising, changing and changing and giving the actors ideas about what I'd like to see them do. So, yeah. But I don't think, "Boy, I want to have something shocking here." In fact, I try not to think at all. I've spent my life reading. I've read everything. And I'm convinced I'm reading so that I can finally say, "Well, you know, I've read it all and I can throw it all out be-

Originally published in *Village Voice*, August 10–16, 1994, 29–34 (excerpt).

cause none of it works." We're both working from circumstance. You've said in the past that your whole artistic life is an interplay between what other people—members of your company, the NEA, Peter Sellars, me, anybody—what they ask you to do and what you really want to do. Well, I feel other people are telling me what to do all the time through books, through encounters, and I want to be open to all of that.

EL: But you have a different experience, I think, because you write your own material. So you have to synthesize what comes into you at an earlier point than I do.

RF: No, that's why I think I'm a better writer than a director. I think I'm a very conventional director who ends up taking too much of the wildness out of the texts I start with. I've always felt somehow I don't have the freedom and courage in front of my actors to appear as spaced out as I allow myself to be when I'm writing.

EL: That's really funny because I feel like I'm totally spaced out as a director and I'm embarrassed I'm so spaced out.

RF: I always hope that I can dare to be that way.

EL: Your things are wilder, if you want to talk wild, they're wilder—your images—than my images are wild. You know what I mean? I am, as a director, not as clear as you are, having worked with you I know I'm just not as clear in the moment and I let things pass by for a long time that you would have manipulated. Yours is a flow of things that you manipulate. I let things pass by that I don't like, and leave them there for long periods of time because I don't have any impulse to manipulate them.

RF: But somehow your structure, the structure of your staging, ends up seeming closer to the disassociative techniques of my writing and lot of advanced twentieth-century writing.

EL: Oh, I would never had said that. Never.

RF: Really? Because your pieces, especially lately, are built so much on this strategy of interruption.

EL: Yeah, but I see yours as interruption. I see yours as constant interruption. But you use language to smooth over the interruption. You know you make the transition in the language. I see it as a constant interruption of an idea by a shifting of meaning.

RF: Well that's what I try to do. But I still feel, watching one of my shows, that I've lost nerve in staging it. I wish my staging could be as open to decentered impulses as the staging of your pieces.

EL: And so much of the staging in my work is because I can't make an action. Because I can't do something.

RF: But every artist finds out that at a certain point mastery can be a prison, a deadening habit. You can free yourself by exploiting and glorifying your limitations. I mean that's what it's all about. Otherwise you're an entertainer. You know how to do things that other people have learned how to do, and you learn to do the same thing just to please.

EL: Don't you ever use that, the sense that you are an entertainer? When people ask me what I do, I tell them I'm an entertainer. I'm a little confused about what entertainment actually is.

RF: Well, that's because everybody is different, every audience is different.

EL: When I said that, I didn't say it to shock. I used to say things to shock because I took so many lessons from Richard Schechner that way—to get people's attention. I knew I had to draw attention. In Europe there's still for me—you probably know this better than I do—there's still this idea that there's art and then there's popular culture. They are still very separate even though there are some forays to combine the two. But for me the two are the same. Probably the difference between my work and someone else's is that I have more control and there is not so much money involved. Period. But the things I am dealing with are the same as popular culture. The same themes, the same ideas.

RF: I think there is a difference but I want both.

EL: Well, what is the difference?

RF: I think that art is attempting to evoke something that you are not yet. And I think entertainment or popular culture is talking to that person that you are now. In other words, most of my experiences of art are difficult experiences. It's like going to the gym, which I don't like to do. But if I do go to the gym and I work out, I feel better afterward. Life is better. And entertainment is just like somebody stroking you. Now, I'm not putting down either one. I'm fallible enough so that I want both. I want entertainment all the time because I'm a lazy slob and I want to be stroked. But I know that if I can force myself to go to the gym of art my whole body would be toned up. My whole mind would be toned up!

EL: You see, I just hate that. And I hate it because it reminds me of church. I hate it. And I think it's my laziness. I do have a real deep laziness. And I don't want to be thought of as someone who teaches. If I talked that way, I'd suddenly feel like, "Oh, my God, that places me above other people in some way" . . . that I have something to teach them.

RF: Ah, but the only person I want to teach is myself because I know I'm a stupid slob.

EL: But then, why are you doing it in front of people?

RF: Ah . . . in my case it's because I'm an asocial person, and I discovered the theater was a way to have contact with other human beings, which I still need.

EL: I think that's where we come together. Yeah, I think that if I had stayed in painting, which was where I began, I would be a hermit now. I would see no one and speak to no one. All of my connections to the world come through this extremely social world of theater.

RF: And how did you get involved with the theater?

EL: I was painting alone in my garret room, and on the weekends for entertainment I became involved in a theater company in town. But it was not serious. Theater was not a serious endeavor, not an artistic endeavor in any way.

RF: But I don't understand. Because if you say that you would have become a hermit . . . then there was a drive to go out. There was a glamour about the theater that must have attracted you.

EL: Sure. And there was a rebellion too. Because my teachers would say, "Don't spend your time down there. That's trash." Which made me want to do it more. So I always had that kind of push-pull one way or another . . . If I had been a man I would have gone on to Yale, to graduate school in painting. And even now the thought is horrifying.

RF: Well, I went to graduate school at Yale, in theater, but that was only to stay out of the army. But I was involved in the theater much earlier than you. I got taken, as a kid, to see Gilbert and Sullivan. I remember, first, the glamour of the footlights. And then the orchestra was playing and the curtain was going up on a painted backdrop of this other world. Those two moments just seemed to be an escape into fairyland. And as I got deeper into the theater I began to realize that it was all substanceless, and I began to try to fill the theater with material from various other arts. Even when I was very young I was interested in anything that seemed nightmarish or strange or weird. And trying to figure out what made something weird led me into the effort to analyze it, which led me deeper into literature and philosophy. I was so shaken and thrilled in a scary way by these things I was attracted to that I, a very conventional, shy young man, wanted to make them safe for myself by being able to understand them. That's still what I do. Allowing impulses that seem hostile, frightening, and scary to surface, and then trying to understand them by staging them. The director tries to understand the crazy writer.

EL: That's trying to understand yourself. Me, I always put it at a slight distance and say I'm trying to understand the writer, or I'm trying to understand the person onstage, the performer, so I'm further behind them. I always say when people ask if it's about me, no, it's coming through me. It's not about me. To hide, I suppose.

RF: It sounds like both of us get the energy to make theater because of our fight with the theater.

EL: Yes. And it started so strongly with Schechner. The fight with the idea of what a performer has to feel when he's performing. And in that moment in *Pain(T)* when I saw the two women under the table fighting, and I thought, "My god, they're not really fighting, they're not even thinking about really fighting." Yet it evoked in me everything about that struggle. Even though I know intellectually that that kind of thing was possible in the theater, and I had been trying to find it, with Schechner, working against him, who always wanted the performer to feel the thing itself, though he never said it. That was his major focus. I remember *Pain(T)* was the first time I felt it in my gut so strongly.

RF: Felt so strongly what?

EL: That you could get the feeling inside you without seeing a person truly being that. Without having to look out a window and seeing a person actually being beaten up. You could get the feeling in your stomach of what that kind of violence was, through a series of actions that were in essence abstract.

RF: Well, the main thing I do think about is that everything I do onstage, everything that can be written, is a lie, because it's only a part of the truth. So my life is an attempt, in all different realms, to continually do something that says, for instance, "It is a fact that I hate you. Yes *but* then again—!" Or, "It is a fact that I love you. . . . Yes—*but* then again . . . ." And that "Yes *but*" is what I'm most deeply concerned with. Does that concern of mine relate in any way to the fact that your own productions do seem structured around a series of interruptions? Are these interruptions in order to negate, or in order to put in quotes what just happened? Or is there another reason? Or is it just that somehow it gives you energy, just turns you on?

EL: I don't know. . . . I can't think really specifically of something that didn't need to be interrupted for me and that I interrupted. It's obviously so deep that I don't even experience it as an interruption, just as I don't experience those other things as abstract. Maybe that's why I've made a whole kind of theory that I don't like to see anything alone on the stage. If there's one thing happening on the stage I have to see at least two other counterpoints to it.

RF: *[Laughter]* Me too.

EL: At least two. And if I don't then I get bored. I can't even deal with listening to people if we have to read a script to begin with—I can't even deal with listening to one person after another, reading the script without putting something over here that I can look at at the same time, to go back and forth between.

RF: So we both need at least two things happening at once. I think that's what, in a nutshell, distinguishes us, and people like us, from other theater. Because when I used to do shows for Joe Papp, he used to be very smart about saying, in a kind of Ibsenite way, "You know, Richard, this one moment in your production is distracting us from the main thrust of where you're trying to go." And I realized at a certain moment that, of course, it's only the distractions, it's only the suggestion that life goes off in a million different directions at all moments, that provide, for me, an interesting subject for art. That defines the real antitheatrical tradition, which you and I are into, as opposed to that other Ibsenite tradition that still demands you craft the audience's attention toward that specific "important" thing that is supposed to happen inside your preconceived premise. And once you've thrown out that notion, you're in a new theatrical world.

. . . . .

RF: Every good thing we do comes from problem solving.

EL: Yeah.

RF: In the early days I was using nonactors and they couldn't talk loudly enough, or properly, so I put their voices on tape so they could be altered.

EL: Well, what is your problem solving now? I'm talking about staging, not writing.

RF: OK . . . *[Long pause]* Well, my big problem is how to stage things so that what becomes exciting is following the way the language of each line tries to say something but stumbles a little and implies something else, and the character's response to that is to decide whether to respond to the interruption or to the line.

EL: But that's there in your head, not on the stage. That's in there *[pointing to her head]*.

RF: Exactly. But my problem, what fascinates me, is how to make that interesting and theatrically involving on the stage.

EL: That's a big subject. Let's get down to, OK, you have this line that you're working on and you have a performer there and somehow when you hear the performer say the line, you don't see the trajectory, you don't see the way it resolves or doesn't resolve. What do you do?

RF: I do the same thing that you do. Calculated interruption. Maybe I have someone, in the middle of the line, enter carrying a bouquet of flowers.

EL: You distract them to the other side of the stage. *[Laughter]* Oh, wicked!

RF: But of course. It's all a series of distractions from what you think you should be following, which is always the wrong thing to follow. . . .

EL: But don't you think that it's that old thing that we always argue about, about language? Somehow I don't have a hierarchical idea of where language sits within ways of communicating, and I think you do. Mine is not an intellectual thing. It's just that I didn't grow up with that kind of education.

RF: I suppose I do, though I want to use language to escape language.

EL: Yeah, but I've already escaped language.

RF: Well, I feel trapped.

EL: It's not that I feel trapped, I find it *is* a trap, that I have to constantly move around. But language to me is like what for a child the color red is. I don't have any associations of its power. . . . No, that's wrong too. I don't look to it for anything but entertainment.

RF: I feel ruled by it. I look at language as if it were a kind of Ping-Pong game in which there are a lot of little balls hitting things and going off in strange trajectories and you're dominated, your life is ruled by the fact that these things are accidentally hitting this way, that way, that way. And I want to figure out the scheme of that so I can be clear of it.

EL: And that's what you write about.

RF: I think the reason is—and this should be a lesson to us, we never learn our lessons—that at least at this point in the twentieth century when we're making the kind of art we're both making . . . [our] task is to dredge up everything rejected, everything that isn't allowed, and discover that energy and beauty are functioning in those things. . . . You could make the case that this perverse historical period we're in produces serious art only if it's perverse. And I'd like to think that I am forced into what I know is a perverse strategy by the times. I'd like to think that in happier, healthier times maybe I wouldn't even be an artist.

EL: Yeah, I understand that a little bit now. You've said that a lot to me, but I haven't really understood it till recently. I've had this feeling of not being an artist. I don't know what it means to me. But I've had at least the idea in my mind that it's possible not to be that, not to do that. Maybe it's age. I've had a vision of just doing landscape architecture. It has to do with figuring out how to replant the earth the way it was. Returning it. You know, some obsessive thing like that. Returning it to the way it might have been naturally.

RF: I've reached a point where I'm not sure I could give up writing. But I could give up directing, I think. Actually, ideally, I'd like to find another director and I would just go sit at rehearsals as the playwright. I was realizing just the other day that practically every moment that I'm conscious, I have the urge to say, "Wait a minute—This life that is passing through me, I want it to be more jewel-like." What I mean is that I don't want things coming in and passing through my head the way they are doing now. I want there to be other surfaces inside me that they bounce off of—like light bounces around inside a jewel. So a new structure is made by that bouncing around. And that's why I have to write, to evoke that, to turn myself over to that imagined "thing."

EL: The closest I come to that is landscape architecture. I want to organize space. I can't think unless I'm organizing space. Now obviously I've thought, "Oh, I'll go outside." I realize now, that's a big change. I'd no longer be an artist. I'd be somebody organizing landscape, which is not being an artist. But it's the same. Yours with words, mine with space.

RF: I don't see the difference between doing that and what I think many contemporary artists do. Just messing around with materials until you find what turns you on, what gives you a thrill.

EL: Yeah, but I always have in the back of my mind these people who will be sitting and watching. And I know when I'm messing around and I don't care that they're there—and I know when I'm messing around and I do care.

RF: Because I've always thought, perversely again, that my moral task in life was to dare to show more and more of the messing around that just turned me on. Without caring what the response is.

EL: Oh yes. Me too.

RF: I do care. But that's a failure on my part.

EL: That's right, yes. And I've always felt that way too.

Evidence (1972). Ontological-Hysteric Theater, Theater for a New City, NYC. *From left:* Jessica Harper, George McGrath, Bob Schlee. Photograph copyright © 1972, Babette Mangolte, all rights of reproduction reserved.

Particle Theory (1973). Ontological-Hysteric Theater, Theater for a New City, NYC. Photograph copyright © 1973, Babette Mangolte, all rights of reproduction reserved.

Sophia = (Wisdom): Part 3: The Cliffs (1973). Ontological-Hysteric Theater, Theater for a New City, NYC. Kate Manheim, left center, with covered eyes. Photograph copyright © 1973, Babette Mangolte, all rights of reproduction reserved.

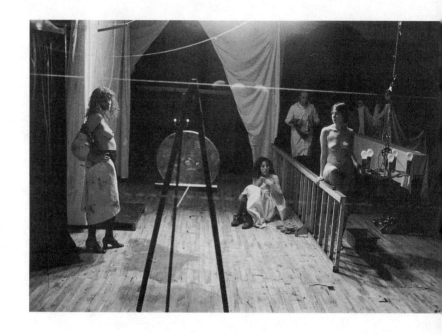

Pain(T) (1974). Ontological-Hysteric Theater, 141 Wooster Street, NYC. *From left:* Kate Manheim, Nora Manheim, Mimi Johnson. Photograph copyright © 1974, Babette Mangolte, all rights of reproduction reserved.

Pandering to the Masses: A Misrepresentation (1975). Ontological-Hysteric Theater, 491 Broadway. *From left:* Bob Fleischer, Kate Manheim. Photograph copyright © 1975, Babette Mangolte, all rights of reproduction reserved.

Rhoda in Potatoland (1975–76). Ontological-Hysteric Theater, 491 Broadway. Photograph copyright © 1975, Babette Mangolte, all rights of reproduction reserved.

Threepenny Opera (1976). Vivian Beaumont Theatre at Lincoln Center. Photograph copyright © 1976, Babette Mangolte, all rights of reproduction reserved.

Book of Splendors, Part Two (1977). Ontological-Hysteric Theater, 491 Broadway. *From left:* Cynthia Gillette, Kate Manheim. Photograph copyright © 1977, Babette Mangolte, all rights of reproduction reserved.

Blvd de Paris: I've Got the Shakes (1977). Ontological-Hysteric
Theater, 491 Broadway. NYC. Photograph copyright © 1977,
Babette Mangolte, all rights of reproduction reserved.

Strong Medicine (1978–79). Feature film, produced by the Ontological-Hysteric Theater. Kate Manheim, Gerald Rabkin. Photograph copyright © 1978, Babette Mangolte, all rights of reproduction reserved.

The American Imagination (1983). Music-Theatre Group, NYC.
Photograph copyright © 1983, Babette Mangolte, all rights of
reproduction reserved.

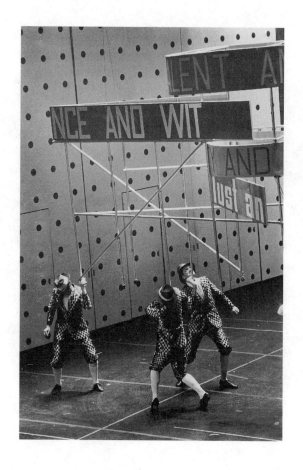

The Birth of a Poet (1985). Brooklyn Academy of Music. Photograph by Paula Court.

Miss Universal Happiness (1985). Ontological-Hysteric Theater and the Wooster Group, Performing Garage, NYC. *From left:* Steve Buscemi, Kate Valk, Willem Dafoe, Ron Vawter, Elizabeth LeCompte. Photograph by Paula Court.

The Cure (1986). Performing Garage, NYC. Kate Manheim.
Photograph by Pamela Duffy.

Lava (1989). Ontological-Hysteric Theater and the Wooster Group,
Performing Garage. *From left:* Neil Bradley, Kyle de Camp,
Matthew Courtney, Peter Davis. Photograph by Paula Court.

Eddie Goes to Poetry City (Part 2) (1991). La MaMa Annex, NYC.
From left: Kyle de Camp, Henry Stram. Photograph by Paula Court.

The Mind King (1992). Ontological-Hysteric Theater at St. Mark's Church in-the-Bowery. *From top:* David Patrick Kelly, Henry Stram. Photograph by Paula Court.

My Head Was a Sledgehammer (1994). Ontological-Hysteric Theater at St. Mark's Church in-the-Bowery. *From left:* Thomas Jay Ryan, Henry Stram. Photograph by Paula Court.

Benita Canova (1997). Ontological-Hysteric Theater at St. Mark's Church in-the-Bowery. *From left:* Joanna P. Adler, David Greenspan. Photograph by Paula Court.

IV Writings

Richard Foreman

Ontological-Hysteric Manifesto I

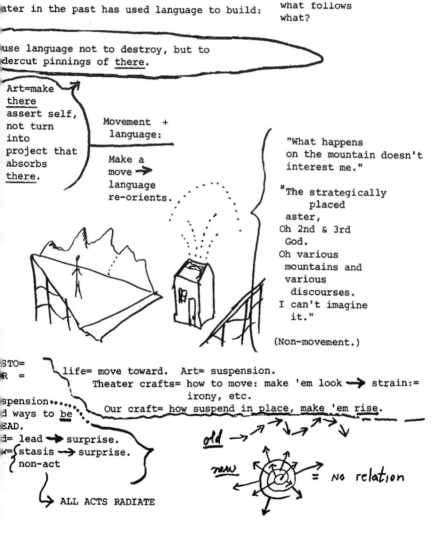

...ater in the past has used language to build: what follows what?

...use language not to destroy, but to ...dercut pinnings of <u>there</u>.

Art=make <u>there</u> assert self, not turn into project that absorbs <u>there</u>.

Movement + language:

Make a move → language re-orients.

"What happens on the mountain doesn't interest me."

"The strategically placed aster, Oh 2nd & 3rd God. Oh various mountains and various discourses. I can't imagine it."

(Non-movement.)

STO= R = ...spension... ...d ways to <u>be</u> ...AD.

...d= lead → surprise.
...w=⎰stasis → surprise.
   ⎱non-act

↳ ALL ACTS RADIATE

life= move toward. Art= suspension.
Theater crafts= how to move: make 'em look → strain:= irony, etc.
Our craft= how suspend in <u>place</u>, make 'em <u>rise</u>.

<u>old</u> →
<u>new</u>   = No relation

Originally published in *Performance* 1, no. 2 (April 1972): 61–85. Reprinted in Richard Foreman, *Plays and Manifestos,* ed. by Kate Davy (New York: New York Univ. Press, 1976).

THEATER

The stage. Destroy it carefully, not with effort but with delicate maneuvers.

Why? Heavy destruction vs. light destruction.
What distorts is excellent.
What distorts with its weight.

Distortions: 1) logic -- as in realism, which we reject because the mind already "knows" the next move and so is not alive to that next move.
2) chance & accident & the arbitrary -- which we reject because within too short a time each choice so determined becomes equally predictable as "item produced by chance, accident, etc."
3) the new possibility (what distorts with its weight) -- a subtle insertion between logic and accident, which keeps the mind alive as it evades over-quick integration into the mental system. CHOOSE THIS ALWAYS

The field of the play is distorted by the objects within the play, so that each object distorts each other object and the mental pre-set is excluded.

MANIFESTO !

»lem of art-- how
»ake people watch
right <u>thing</u> while
going on... **SO** ⟶ Not watch changing
relations:
But watch what
<u>doesn't</u> change in
midst of it.....

---

nguage
»jects us
orien-
:ion
ısage." --
:kham:
ıt's
».

Art: not concerned with essence
But with THING
used in such a way
that it vanishes
& what is
left is suspension:
In life. ⟶ .thing is tool --we get
somewhere.
In art. ⟷ never get there
Suspended.
Why? Create a ZONE
in which placed
things (head) luminate!

1967-- Suddenly the theater seems ridiculous in <u>all</u> its
manifestations and continues to do so in 1971. I.e.,
Peter Brook staged <u>Midsummer Night's Dream</u>. The actors
enter onstage and immediately, the absurdity-- both
in the orchestrated speech and activity-- as Stella,
Judd, et al. realized several years ago...one must reject
composition in favor of shape (or something else)...
Why? Because the resonance must be between the head
and the object. The resonance between the elements
of the object is now a DEAD THING.

1971--Lenox, summer. I sit, at sunrise, and stare out
into the trees, listening to the birds-- i.e. 100
invisible birds in counterpoint. My head, savoring
that interweaving of themes, performs in a good way--
performs in the way that heretofore I have felt art
should make it perform. But suddenly (drama!)
that often-before entertained notion crystallizes in
my head in such a way that a chapter ends, the book
closes, and I have no more interest (no more risk,
no more "unknown") in such an art based on counterpoint & relationship. What can replace it? Don't
know.... The painters have discovered "shape." What
can the theater discover?

I.E.

Only one theatrical problem exists now: How to
create a stage performance in which the spectator
experiences the danger of art not as involvement or
risk or excitement, not as something that reaches out
to vulnerable areas of his person,
      but rather
the danger as a possible <u>decision</u> he (spectator) may
make upon the occasion of confronting the work of art.
The work of art as a <u>contest</u> between object (or
process) and viewer. ⌈Old notions of drama (up thru
Grotowski-Brook-Chaikin)= the danger of circumstance
turning in such a way that we are "trapped" in an
emotional commitment of one sort or another.⌋
The new ontological mode of theater (within which
hysteria lies as a seed/spark which forces the
unseeable to cast shadows) --
The ontological-hysteric theater: the danger that
arises when one chooses to climb a mountain and-- half-
way up-- wishes one hadn't.

> Art till now= appealingness: Making an object
> that we fall in love with. Make the obsessional
> object.

     NOT ART/No more art, naturally. Yet the aesthetic
thrill. The point is, of course, that "art" no longer
provides (provokes?) the aesthetic thrill. In a world
of scarcity (now psychically superceded if not yet
practically) the one was against the other.
Conflict at root of drama. OK. It's all so simple,
really. Now-- art can't be based in conflict. Old art
aroused, empathized with that, made our inner nature
vibrate to that in such a way that it was "profound."
The grounds of conflict are now seen as...not between
entities, but within the single unitary occasion which
could exist-- could not exist. That oscillation
replacing "conflict."

where oscillates the conflict ?
conflict o —→ ←— o
conflict ←— • —→
• • • • • • • • ?

But there is no center, the conflict is between the idea
of a center and the idea of a field. (The idea of a
center= old-fashioned "being"; the idea of a field=
old-fashioned "not-being.")

A FIELD BY ITS NATURE CONTAINS ONLY HARMONY.
In our attempt to hold together a center, we mistakenly
   view the field perhaps as one in which particles,
   etc., form a kind of conflict situation. But:
Not true.
Conflict there, as elsewhere, an illusion. (Absence
   of conflict doesn't mean absence of dynamics.
   Conflict BLOCKS dynamics.
Ecstasy = all forces operate at once to produce STASIS!
            (Replace conflict-- push and pull of selected
                 forces-- with total action of all forces.
                 That is stasis, that is ecstasy.)

Wittgenstein:
If mean= intend. Anything is intended.
Any intend.
Use anything, to mean anything: but, the system must
      have a rigor.
Mean something by a movement·of the hand-- was it the
      movement that he meant?

To express something which can only be expressed by this
      movement.

To read off the "said" from the face of the thought?
No-- our theater is making harmony. Singing counterpoint
      in language-- swimming in language in a way
      appropriate to the ongoing internal (mental) activity..

So : language systems:
            THUD!
            (Start out speaking in own terms,
            system created in terms of play
            by using own concerns!)

$$\left(\begin{array}{l} \text{think of swimming, think of singing,} \\ \quad \text{think of the picnic, think of the grass} \\ \qquad \text{glass} \\ \qquad \text{glass} \\ \qquad \text{glass.} \\ \text{(Has a system begun to be created?)} \end{array} \right)$$

Now:

Acting against materials (the table, the floor, the other actor's body) is establishing this new language that doesn't <u>read</u> but "illustrates." (I.e., thinking against things.)   Pick proper interference.
Like new <u>motor</u>.

$$\text{MOTOR} + \left/ \begin{array}{l} \underline{slight \ shifts} \\ \underline{stuckness} ? \longrightarrow glue: \end{array} \right.$$

ALWAYS NOTATE YOUR EXACT SITUATION AND PROCESS WHEN <u>WRITING</u>!

TAKE TWO RULES CONTRADICTORY IN NATURE. FOLLOWING BOTH MEANS SUCCESS.

In 1968, the theater became hopeless. I suppose the immediate revulsion is always against the artificiality or something related to that, although artificiality itself is noble enough-- being the HUMAN contribution which, if properly posited, lifts the moment...turns nature herself into a construct of delight.

Ah-- that is the point is it not? To make a construct, which must be the motive behind all art effort. So where does the theater's artificiality turn sour? At what point does the "construct" give way to the lie, to the exaggeration? That may be the point: to isolate the difference between exaggeration and invention. Whereas exaggeration destroys balance, and invention is constantly replacing the center of harmony, shifting it slightly in such a way that the shift, the moment of shift, the act of the shift, becomes-- if experienced as the specific OCCURRING EVENT that it is-- an occasion for testing oneself, as climbing a mountain is a test of the body.

Art, of course, tests the soul, tests the psyche. That is to say-- purely a matter of vibrations. Now, where do these vibrations vibrate? What fluid is it in which the resonating wave patterns are established?

Well, folks-- ! The vibrations are in the head, of course. And they are most certainly produced by the (demonstrable) scanning mechanism of the brain. And the universe-- which exists to us as a direct "production" of that scanning and in-the-instant-rescanning, must enter into a new relationship to the art work. No longer the relation between a changing world (events march on) and a posited ego which VIEWS events and in so doing EXTRACTS art from the flux of the world-- while in that EXTRACTION lies the terror that manifests itself as "conflict and expression" in drama.....but a (new) relationship in which the world is essentially a repeating mechanism (which it is on both its building block level and its higher cyclic levels) and the scanning mechanism superimposed on the repeating mechanism slowly builds an edifice. (The way nature and history build.)

Two DIFFERENT kinds of edifice are built, however. One of them is called "life" (in the private sense of "I have lived a life") ...and the other shall be called art, though this "art" is clearly something different from what has been called art up to this point.

For this new "art"-- perhaps we should not think of it as an edifice but as an accretion, as deposited sludge-- this new art is not EXTRACTED from the flux of life, and is therefore in no sense a mirror or representation-- but a parallel phenomenon to life itself. The scanning mechanism produces the <u>lived experience</u> when it is passive. I.e., the input rhythms are dominant; and the scanning mechanism produces <u>art</u> when <u>it</u> is the ACTIVE element, when <u>its</u> rhythms dominate the scanned object. (The actual "making" of the art object then becomes essentially a matter of notation.)

So hopefully, we end up with a new art that serves two essential, related functions:

1) Evidence: useful as example to others, of the harmony that results from an awareness and conscious employment of our mechanism which is our "self" in its properly industrious way upon the world (that flux of "everything that is the case").

Evidence....to give courage to ourself and others to be alive from moment to moment, which means to accept both flux (presentation and representation to consciousness as reality) and an INTERSECTING process--scanning--which is the perpetual constituting and reconstituting of the self. The new work of art-as-evidence leaves a <u>tracing</u> <u>in</u> <u>matter</u> of this intersecting, and encourages a courageous "tuning" of the old self to the new awareness.

2) ORDEAL. The artistic experience <u>must</u> be an ordeal to be undergone. The rhythms <u>must</u> be in a certain way difficult and uncongenial. Uncongenial elements are then redeemed by a clarity in the moment-to-moment, smallest unit of progression. After all-- clarity is relatively easy (at least the "feeling" of clarity) in terms of large structures because simplification can always be wrought on a large structure (simplification often being the bastardization of clarity).

But CLARITY is so difficult in the smallest steps from one moment to the next, because on the miniscule level, clarity is muddled either by the "logic" of progression (which is really a form of sleepwalking) or by the predictability of the opposite choice-- the surreal-absurdist choice of the arbitrary & accidental & haphazard step.

Of course
 ORDEAL
is the only experience that remains. And clarity is the mode in which the ordeal becomes ecstatic.

Art is not beauty of
description or depth of
emotion, it is making a
machine, not to do some-
thing to audience, but that
makes itself run on new
fuel. Can this machine
run? Most machines (art)
run on audience fuel--
(Man's piggish desire to
be at the center, to be
made to feel there is
"caringness" built into
the world: old art
manipulates that, tries to
get a response: fuel is
DESIRE in that case.
FIND FUEL OTHER THAN
DESIRE! Nervous energy?
Basic hum of life?
Vibration?) (Desire
kills vibration, gets
too crude)

WE MAKE A PERPETUAL
MOTION MACHINE. (The
closer to that ideal the
better. Run on less and
less fuel...that's the
goal of the new art
machine.)

I REPEAT !

I want to be <u>seized</u> by the elusive, unexpected aliveness
of the moment.
Surprise at the center: not the surprise of the least-
expected.....because that (least-expected) is a reaction
that "places" it and makes it no longer elusive. But
surprised by
a freshness
of moment that eludes
                    constantly refreshes. You go toward it
and can't seize it? You don't go toward it............

Art to me=
energy of wa
to know (ale
without desi
move off the
center, off
energy itsel
the object.
happy <u>NOT</u> kn
in condition
wanting to k
Be joyous in
tension.

Most art is
created by
people trying to
make their idea,
emotion, thing-
imagined, <u>be-there</u>
<u>more</u>. They re-
inforce. I want
my imagined to be an

Write by thi
<u>against</u> the
material. S
you don't wa
to <u>convince</u>
of your visi
etc.-- but t
it be inform
the disinteg
now-moment.

occasion wherein the not-imagined-by-me can be there.
My work= to deny my assertion (imagined) is true (is
there).
    by letting moment disintegrate it, as no assertion
really true in the face of the elusive now, the real
moment, which in its bottomlessness turns what it holds
into the bottomless anti-matter of what is <u>itself</u> in
the rigidity and deadness of before and after.

Subject of theater-- vanity: in all: nothing real
or of any great matter, including <u>that</u> fact: So it is
ALL THEATER.

1) Used to be--   like a staggered race
                     Relations (beauty)
                     Now that's a cliché.
        So-- no relations:
             But shape?   relates to head

             Head: keep dealing with throb in head.

2) Undercut
        Set up irritant
             against line of the scene.
                  (Bright lights?)
                       intersect with other realm.

Don't sustain anything

1) Erotic ~~angel~~ angel : —a shape

Subject: THE EFFORT OF PUTTING WHATEVER ARISES TOGETHER
         That <u>EFFORT</u>.

Subject: Make everything dumb enough to allow what is
         really happening to happen.

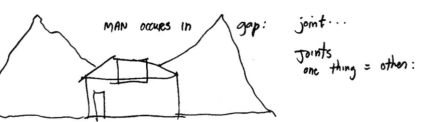

MAN occurs in   gap: joint...

Joints one thing = other:

"a new cadence means a new idea."

only: → *get* rhythm of the mind as something that acts
vis-a-vis entering dots: which leave _traces_ that other
dots bounce off. So mind is input folded over imput.
GET THAT!

Manifesto!

IN which
everything
achieves a
final form.

BEN: (That writing) says "I gave up writing plays
three years ago." (_Pause._) The fat lamp descends.
That is not written but imagined. (_Pause._)
Come into the writer's workshop.

A second lamp down, hanging over the first. A hum.

Get that second lamp outa here.

LEGEND: "WHY?"

BEN: Nobody has a right to ask me questions who doesn't
show himself.

Crew comes and closes curtains on screen on which title
is projected.

Why.

Thud, pause.

The minute I formed that word carefully it was
an imitation.

The curtain reopens by itself. A slide of an ancient
auto is projected on the screen. Pause. Then the same
picture is projected on the table-top.

VOICE: One picture must not be allowed to view the
other picture.

Music begins. A sign comes down--"Cousins in
photography." The music stops. Ben has exited. He
returns with a rope, throws it over the screen, with a
hook on the end of the rope, and starts to pull.

BEN: Oh well. (Pause.) Make something.
ALL: Can you describe it, Ben?

Program Note for *Particle Theory*

Here's what the theater should do—never try to express but, rather, make the object be more itself. (Object = whatever we stumble upon without humanist pre-conceptions.) I.e., delineate the object's (act's, item's, event's) boundaries ever more clearly. To make an item "express" something is to allow other, known ingredients to bleed into and through it. Result: expression is always a distortion of what-is. Result: to the degree that art expresses (distorts), it is Romantic art which gives nothing but exotic experience. Believe it or not, what I am interested in is a classical art which tries to clean the perceptual lens, to notate what is really there in the consciousness when it (abrasive experience of perception) stumbles over what gets in the way at the first and each succeeding moment of "effort" (i.e., living).

Here is the audience, waiting for something to happen. They have their own lives, of course. In the context of their own lives, do they wait for something to happen? Well, in their own lives they are waiting for something to happen whenever they are dissatisfied, that is—when they have too much of a certain other thing, which is called "nothing interesting happening." Then, they feel the lack of "nothing much happening"'s opposite—that opposite being called adventure or activity—or something-exciting-happening. If, however, the balance between not-much-happening and exciting-happening is O.K., then they feel O.K. and they are not waiting for something to happen.

We go to escapist, exotic entertainment out of a lack of eventfulness in our lives for which we want to compensate. But that is not the root of

Unpublished, 1973.

the need for art—though of course the theater rarely serves the need for art, usually the need for entertainment (i.e., exoticism and adventure). We go to art (theater) because we have too much of the not-art in our lives and we want to restore the balance with some art—we ache for need of what art does to us. Now, what does art do to us and how is it related to what theater (both art and entertainment styles) can do to us? What is theater?

Watching. Watching a spectacle. Watching. In life, we have too much not-watching, either because there is nothing (interesting) to watch (nothing's happening) or because it's happening but we don't watch it—we are in it, involved, and we don't have time to watch also.

We are in it and it gets watched, but not through our conscious watching. The need for conscious watching is the need for art. The body passive, but the watching conscious, active: getting a chance to do conscious watching. In life—perceptions enter, click, click, automatically.

We have a need (if we need art) to stop that automatic watching, because perception allowed into us that way ends by making us feel fat, satiated, depressed, flabby in the consciousness. We lose the sense of ourselves as the perceiving entity, and with the loss of THAT sense of the self comes depression and a lack of energy. So, go to the art (theater) to regain that sense of the self as a perceiving (by choice, consciously) entity.

So, in staging the play—everything is oriented to serving that need. Not to create a convincing exoticism that may be watched (swallowed) unconsciously as we swallow daily life—but to make a machine out of the text, a machine which will challenge the starving perceiving-self . . . challenge it to the self-conscious, active watching it needs in order to restore human equilibrium to the depressed and exhausted self-system which has been crippled through a deprivation of "consciously perceiving" experiences.

How to Write a Play

(in which i am really telling myself

how, but if you are the right one

i am telling you how, too)

make a kind of beauty that isn't an
ALTERNATIVE to a certain environment
(beauty, adventure, romance, dream, drama all
take you out of your real world and into their
own in the hope you'll return refreshed, wiser,
more compassionate, etc.)
                              but rather
          makes GAPS in the non-beautiful, or look carefully at the
structure of the non-beautiful, whatever it is (and remember that structure is always a combination of the
                        THING
                      and the
                  PERCEIVING of it)
and see where there are small points, gaps, unarticulated or un-mapped
places within it
          (the non-beautiful)
which un-mapped places must be the very places where beauty CAN
be planted in the midst of the heretofore unbeautiful.
          Because the mind's PROJECTED beauty (which is the only
          beauty) . . . can either find itself in the already beautiful (so
          agreed upon) or it
          can MAKE
          Conquer new territories.
              But: while in the midst of the heretofore still un-redeemed
          "non-beautiful" the projection of the will-to-beauty can either

Originally published in *Performing Arts Journal* 1 (fall 1976): 84–92.

be a pure act of will in which there is a pure, willed reversal of
values

  (which can have great strategic value but creates art that
  DOES tend to "wear out"—not, you understand, a negative
  judgement)
      or
our method.
find the heretofore un-mapped, un-notated crevices
in the not-yet-beautiful landscape (which is a
collaboration between perceiving mind and world)
and widen the gaps
and plant the seed in those gaps
and make those gaps flower . . . and the plant
over-runs the entire landscape.

  What this amounts to is a DECISION
  view non-beautiful material in such a way that was foreground is now background . . . and the desired beauty is then projected, as the creative act, into the midst of the heretofore rejected
(non-beautiful, un-interesting, cliched, etc.).

Delight is delight.
It aims us to whatever it is that delights us.
Can we make a more CONTROLLED use of that energy of
"being-aimed" by willfully choosing to have a certain object be the one
which arouses that delight-energy? ANY object?

  Of course. That's the task—discover how to be in control,
  how to
  CHOOSE.
  which object shall provoke the delight phenomenon

  (and so increase that per-centage of the world we can say
  "yes" to, and thereby gain an inexhaustible fund of "delight-
  fuel")

        Here's how.

Normally, let us assume we are delighted by a sunset
We are not delighted by a corpse.

But if we place the corpse within a certain composition, let us say—we are then delighted by the composition of which the corpse is a part

So—while we are still not delighted by a corpse, we can be delighted by something (made or found) of which the corpse is a part.

The task of art is to find what heretofore does not delight us, and make that part of some kind of composition in such a way that delight results.

Now, the composition need not be a composition in the expected sense, that is, need not be something that is defined or defines the art-work itself—

The composition may be any "context" in which the material is placed. In much art today, for instance, the context-composition is "the inherited history of Western art." So that the reason a minimalist gesture such as a Morris black box is "delightful" is because we understand it as an intelligent next move chosen in the context of an evolving "game" which has been the game (move and countermove) of Western art.

So in the theater, which is always behind the times, one must ask "ah —what can we include in the on-going context composition which heretofore has been de-valued and kept out, etc., etc., and few people in the theater ask that question and do that thing and so the theater is rarely art, and when it is it creates problems for itself since the audience is not an audience interested in art but in entertainment.

Which means, its audience is interested in being delighted by what they already know in themselves as delightful. And their response to the attempt to include NEW material in the composition—material which they heretofore have categorized as non-delightful—their response is generally negative because they have never been trained to be composition perceivers rather than object perceivers. When they look at theater, they use daily-life perceptual modes and so see things, and not patterns and contexts and compositions.

The rallying cry must be—stop making objects that men can worship.
Art shouldn't add new objects to the world to enslave men. It should begin the process of freeing men by calling into doubt the solidity of objects—and laying bare the fact that it is a web of relations that exists, only; that web held taut in each instance by the focal point of consciousness that is each separate individual consciousness.

In my work, I show the traces of one such web. (The assumption herein is not idealism, because the consciousness is a constructed thing also, on a different level subject to the same laws of configuration as the world outside, a collection of trace elements, not a self-sufficient constituting agent: but the relation between consciousness and "world" is the relation between two intersecting force fields, neither of which is a thing, both of which are a system of relations.)

I show the traces of such web intersections—and by seeing that, you are "reminded" to tune to your own. Find objects in a sense interchangeable (and, in another sense, poignant for that reason). But most of all, find exhilaration and freedom and creative power, for when you see the web of relatedness of all things—which is in a certain ever-alive relation to a "your own web" of consciousness—you then are no longer a blind, hypnotized worshiper of "objects"—but a free man. Capable of self-creation and re-creation in all moments of your life.

Most audiences and critics want to be moved, knocked out. That is a sign of their illness, blindness, need to remain children. Most audiences want a perceivable, nameable content. That is, they want to be able to reduce the experience of the work to a gestalt of some sort that they can carry away from the theater with them.

That means, they want to feel that they have extracted property, capital, from the investment of time in the experience.

NO! The art experience shouldn't ADD to our baggage, that store of images that weighs us down and limits our clear view to the horizons. The art experience should rather (simply) ELIMINATE what keeps us moored to hypnotizing aspects of reality.

Or better—by showing how reality is always a "positive" which is but a response to (an extraction from) a "negative" background, it allows us, in terms of this continual, now revealed polarity, to make contact with the reality that is really-there. Not by social fiat, but by operating at the constituting heart of things.

It is not a matter of getting BEYOND, DEEPER, HIGHER than everyday, normal, agreed on culturally-determined reality, it is a matter of—within the confines of the art experience—allowing ourselves to partake of the "taste" of a perceptive mode that strategically subverts the very OBVIOUS aspects of the gross and childish conditioned perception used to "brow beat us" through life. The gross mode of perception that suppresses the contradiction at the heart of each consciously posited "object."

The artist must search for what has never been seen before.

BUT

Not simply a new "monster." Not a new "that knocks me out like . . ."
(a pyramid, Shakespeare, sex, etc.)

But

a new

object which once found

is hard to see. Maybe it's not even "there."

We live in a world of traces. Things leave traces. We must never try to
make man believe that what is by definition constituted as a "trace," has
indeed a different kind of reality—that of "object."

The emotion must never come, as it usually does, through our being
convinced of the reality of the image or event presented, but only the ecstatic emotion of one's own seeing of things. Delight in one's own energy.

NEVER awe or delight in the "worshipful way" we feel emotion
when we are awed or moved by the "other" which seems like an alien
other in which we "wish" we could partake (all romantic art).

Need for Confrontation

Art   += to CONFRONT the object

Kitsch =   atmosphere replaces object   What is the object?
distance between you and object The encountered object,
de-creased by atmosphere   encountered in making
which makes you FEEL   the work: the "real"
at one with the object   chair, body, word,
because the atmosphere   noise, etc.
is felt to be that exuded
by the object. But then
object and you (feeling)
are one and there is
no ENCOUNTER, and no   The constructed object
seeing. (To play the   end with (art) is the we
subtext, rather than   STRUCTURE of the
the object, for instance.)   articulating process. The
MAKING A THING BE-
THERE AS ITSELF
(in its web of relations).
Process.

The artist doesn't explain, analyze the object . . . but he sets it up so that one CONFRONTS in the realist fashion its BEING-THERE which is a confrontation to your own BEING-THERE.

PARADOX

The way to confront the object is to allow it its own life—let it grow its own shoots in directions that do not re-inforce its being-in-life for use as a tool, but that suggest a compositional scheme not centered on useful human expectations. So, let the chair that is for sitting have a string run from it to an orange, because if chair was just "chair for sitting" we would not "confront" as we not-confront in kitsch because we are too close to the chair, its meaning is too much OUR meaning; but now chair-connected-to-orange is an "alien" chair that we must CONFRONT.

> (To reveal an object or act, gesture, emotion, idea, sound.
> To make it seizable
> To speak its name you must
> make it part of a system not
> its own. Involve it compositionally with
> another realm, which is YOUR realm of pattern
> making isomorphic with your
> mind-process. THEN there is confrontation.)

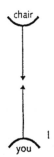

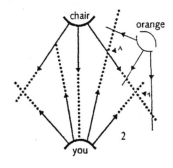

KITSCH
(and sleep) as
the moment of
contact between
you and object
is one dimensional

ART
(and awakedness)
as
the moment of
contact between
you and object

and you are
in a state of
identification
(hypnotized by)
with that face
of itself the
objects presents
to you.

is multifaceted and
often "distant" (point A)
from the object.

also
in 2 (as opposed to 1)
your mind-pattern-
process is being as-it-is:
and THAT structure
inter-acts with the
object structure as-it-is.

In 1 the mind forgets
its own working and there
is no real meeting, only
a 1-dimensional (1 "presented"
face to another) moment of
"official" (cliched)
    "something"
that is too mindless; i.e., one dimensional
(lacking points "A" of 2)
to be a real encounter.

Diagram 2 explains, once and for all, all of my plays!

The message must be "To choose either turbulence or serenity is an error. To choose either knowing or doing is an error." So . . . in the play . . . inject disruption into knowing, and order into passion.

The play is a lecture in which you don't say "This is so . . ." but rather . . . "This occurs to me and it occurs to me that the reason it occurs to me is this act, which occurs to me," and so on and so on, deeper and deeper.

My message is "filling the space with the idea." Free play within the idea. Ability to treat the "field" of the idea as an area for work and discovery. Idea as a field . . . in which something that is not idea (but more physical, sensual, ecstatic) can emerge.

I write to make life handle-able.

The deflection inherent in time. Space—
one makes art to be able to decide what goes into you
and what goes out of you.
To be in control of what goes into you and out
of you is why you decide to make your own art.
Of course . . . as it goes in and out, space and time
give it an uncontrollable "twist."

Lived experience is a certain kind of focus. You focus on an aimed-at while living, and because you are focused on that, you don't see your own gestures. Art: is trying to see your own gestures.

My gesture has always been to pull away, to change what came into me, to make something BETTER, that could then go into me instead of the thing that did not go into me. Hence, to find a way to make better FOOD for myself than was provided by others. My art then (one's art) is a way of being-in-the-world so that the INPUT is the best possible input . . . into me.

Journalism is trying to imitate life. Art is an amplification of the effects encountered in trying to make art.

Art: a machine to effect input. To provide awakening, energy-giving discontinuities. To fight entropy. Art is NOT comment on life. It is fighting the entropy of life-that-seeks equilibrium, that seeks-not-stress, which would lead (as life does) to death. (Inject quantum shocks, discontinuities, to keep twisting us away from sleep, death, into what is "artificially" sustained . . . AWAKENED LIFE, CONSCIOUSNESS!

Form in art—form isn't a container (of content) but rather
a rule for generating a possible "next move."
That's where the subject is (in that next move, dictated or
made possible by the form). The commonly-thought-of content or subject is the pretext to set a process in operation, and
that process is the real subject.

The text is me

It grows like I grow

it extends itself, falls, stumbles over . . . something.

Recovers. It projects itself as it will. Encounters resistance of various sorts, but those resistances turn out to be steps affording a new advance

stretch   extension   twist

stage it: Try to make the compositional aspects be
in relief. The structure as it were. Not the structure
in time, but the structure of the moment.
> (time doesn't exist. It's all **now**
> There's memory-now-future.
> Now doesn't exist
> It's a pivot point
> Make things structured in that pivot point.
> (I.E., frame now, frame not-there)

People who work in time are making things for memory
> Are not clear about here and now
> > Proper analysis here & now
> > > What am I doing now

Man is future-oriented, but life is collision in now between project and what resists it

So: each movement shows what interferes with, contradicts projects—from other levels. Not just conflict of people. What contradicts the "play" itself and its mode-of-being-present.

That privileged object which is the ONE object that must be studied . . . (so that man can study what it is most important he study) . . . that object is not yet "available." Not yet there.

In making a play I am trying to make that important object that is not yet there.

The Carrot and the Stick

My method of procedure in generating texts for performance hasn't really changed in eight years in at least one rather peculiar way. I keep sending myself orders on "how to proceed"—reminders of what I'm aiming at, and piles of these orders accumulate on my desk next to the notebook filled with the scratching that eventually gets shaped into a "play." What I REALLY want to be able to stage some day are these obsessive theoretical out-pourings—but I don't know how, yet . . .

One does not think words, or sentences, or acts, or stories—but only, wherever you are at this minute, waiting to make something—twist, and that twist is, somehow, the unit. And the work is built out of such units.

A certain rhythm of interruption and shifts on a repeating "frame."
(Frames too, alter, but are always frame-like.)
The art . . . must be isomorphic with the feeling aroused by itself. That means, chasing its own tail, which means in turn perpetual motion. The feeling comes after the art which causes the feeling, and yet the art which causes that feeling is made isomorphic with the feeling—and this all conceived *not* as a temporal process. But somehow learn how, in the instant, to shape the moment so that it will be resonant to whatever effects it will produce. Then, when the effect *does* occur, one is truly able to perceive the "structure" of that effect. That's what we should build—models of effect-structures.
Usually, feelings are aroused and out of those feelings one acts. Hate is aroused, murder results. The act *issues* from the emotion. How much better, to discover within the emotion, some sort of framework, along the struts and supports of which one can align one's body, one's imagi-

Originally published in *October* 1 (fall 1976): 84–92.

nation, one's gestures—so that the "act," rather than issuing from the emotion, etches its rather imaginary configuration in the materials of the real world. (The motive, of course, not change, but lucidity. The spiritual motive.)

About four years ago, I discovered how, when suffering from a headache, to lie down on the couch and stop "fighting" the pain, telling the headache to "expand" as it were, until I was alone in a center of a vast web of the throbbing pain—and somehow in that center was a stillness and the pain—no longer resisted—vanished. In the same way—try to generate in the text certain points that are "bad" (whatever that means) in a way that the pain of the headache is bad, and rather than trying to fight to eliminate those points—enter them, let them (the badness) inflate like an entire world in which you can find an entire structure within which a whole life of rigor, passion and intelligence can be lived. The end may be slightly different than the end of "headache elimination," but the starting point is the same. A relaxation and allowing of "bad" material to expand to the very horizons so that *I* am on the inside of *it,* rather than *it* being experienced as a foreign agent within *me.*

Trying to be centered . . . on the circumference. Something inside of you (like a headache, or your response to a "bad" line of dialogue) is a feeling. Relax and let it expand to the horizon, then you are alone at the center. The feeling . . . has become the structure (world) within which you move. Then your movements (your art) indeed become isomorphic (you move along roads laid down by the expanded feeling) . . . with the feelings they, originally, created!

I've always wanted my art to be *about* whatever it was that gave me the energy to make it. My works, therefore, are a mode of literary criticism, in which the object under analysis is itself.

Most literature expresses how the artist feels about a certain sustained "subject." I invariably choose to express how I feel about the preceding moment of generated text. Mostly, how I feel about the energy that generated that preceding moment. Or rather, the relationship between that energy and the one out of many possible ways it chose to crystallize itself. Continual judgments and reflections upon what just was "there." So the critique of the play is not so much built into the play—it is the body and flesh of the play. Indeed, the critique of a play that isn't there—and I feel the play *shouldn't* be there, because if it were there—it would only be there for the *moment* of its performance while what would remain (forever) would be the memory of performance in individual spectators' minds—that memory (selective, judgmental, etc.) immediately a form of critique, and so I chose to make the work out of "what-it-is-that remains" rather than what is momentary (non-existing). So what is articulated and organized is not so much acts, as responses and reflections upon acts.

To understand the work, one should not, of course, ask what it "means," but only—what need does it answer. In my case, the most consistent, passionate need . . . is the need to FILL A SPACE in which I find myself (mentally). That is, I suppose, a kind of erotics of thought . . . using thought to manipulate the imagination, which is a body. Fill that space (where one is now, and then now, and then now) not by being at the center but rather by a twist administered to the imagination-body: an un-natural extention of some sort, generating a new periphery, a difference.

We lack a center, always. By definition (man). It's wrong to try and provide a center (the play should imitate what-it-is to be a self, which is to be centerless). We are peripherally defined creatures. Joy and exhilaration will be attained in the work if it imitates what we really are, which is a process involving a lack at the center which receives a collection of in-mixed traces, so that our mental antenna are constantly feeling out to the "edges" where we imagine those traces to originate. Don't, therefore, think of filling the "space" of the moment, but in the moment, distribute oneself at the periphery. That would be, a union (of "X" with) *other* codes, traces. Then let that union, that in-mixing be the agent that does the act-ing. The "I" doesn't act—the generated sentence, the gesture that results from fold laid back upon fold, the idea that appears as a wrinkle where one line of input stumbles over another—those are the agents of the "act."

My experience continually (life experience, making-art experience) is one of "hummm, that's not quite right" and I try to back away for a new angle of approach, and be seized, there, away from my center, inspired (which means jolted out of line, twisted) by a trace, otherness, irrelevance, "error," which in speaking through me will, as it were, change the rules of the game.

The irony, which is still at the motivating place of the drama, simply attacks different "objects" these days. With the Greeks, it was the irony of an act producing a result opposite to its intention (will effort followed by reversal and revelation). From Shakespeare to Ibsen, the irony was relocated in statements, where a statement is made and can no longer be believed to say what it says, because we know the character is lying, or pretending, or calculating. Now, the irony is in the very *field* of discourse. It pulls the very sentence apart. There is no longer a speaker, towards whom an ironic perspective is to be employed . . . but the total field of words, gestures, acts available to the "speaker"—each "item" in that field is now perceived as ironically meaning its opposite, causing its opposite to "be" the minute it is performed. That is the modern, ironic consciousness. The performing (or naming) "A" evokes (invokes) in that instant, immediately, non-A. It is only against the field non-A that A can make its entry. We KNOW that. It is one of the few things we, in the historical period, know in a more *lived-in* mode of knowing than men of earlier eras.

Poetics of production:
1) A "meaningless" event.
2) A field of experience.
3) A point of view relating 1 to 2.
Think of life as a "music" of these three interpenetrating moments / realms.
The borders between them shifting all the time, of course. An item "A," could shift, oscillating between 1, 2, and perhaps even 3.

        1   2   3
Listen-speak-click of release that's no-mind. Ah!
Learn-create-objective letting-be. Ah!
A possible sequence:

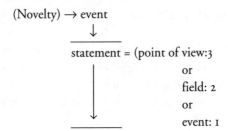

$$(Libido, \quad \overline{\text{The space of}} \atop \text{throwness into}) \rightarrow \begin{array}{c} \text{the statement} \\ \text{then FILLED with} \\ \text{a body-ness} \end{array} \Big\} \quad = 1, 2, \text{ or } 3.$$

$$\downarrow$$

$$\overline{\text{re-statement}} \qquad \text{THIS: a music of} \atop \text{then} \qquad \text{continual counterpoint.} \atop \text{Novelty event.}$$

The most interesting task is to discover the *shape* of the now-moment. So it becomes a matter of forms, more than a matter of structure.

What is the form of the present, and each succeeding present? Then, see what-can-be-done with the form that is the real form of the here-now. Those here-nows as the building blocks of some other structure. But the quality of the blocks determines the possible "style" of the over-all structure.

Now, the form of "now" can be determined only as I try to twist my body (mental) until it FILLS somehow the moment, till it touches the borders of the moment. The meaning then, cannot be in a superim-posed fable, but is in the modes found of being able to inhabit (fill) the spaces of the present, and the sequence of those modes. Meaning is— "how do you live in a space?" Spaces arise, the way mutations are deliv-ered upon the planet—and then life tries to inhabit that new, mutant species. In the attempt to make an arisen space "habitable" (a species is also a "space" for living)—meanings arise, such as "that plant is poison-ous." I am concerned with such meanings.

Meaning? Make an item (the play) that other items can allude to when they are making an effort to crystallize their own meaning-to-them-selves. The play doesn't allude to a real world, through having a "mean-ing." Rather it is there to "give meaning to" anything *else* that wants to take meaning from *it*.

What we need are models for a "way-of-being-in-the-world" that we'd like to remember as a possibility. I'd like, myself, to be "tuned" to the world in the way the play I create is tuned. I establish the world of the play so that hopefully, I can turn to it, and begin resonating to its rhythms.

I generate a text, I make a composition out of what I "know," that is to say—a collection of "meanings" carried around inside me. One meaning . . . in conflict with another meaning. That means, of course, a continually shifting frame of reference. That means of course . . . that there is no conclusion . . . no beginning, middle and end . . . but, intermissions. Until I die. But then I won't be able to write about it.

In the work of art, you are never talking about what you are talking about. You are always using talking-about-subject-"A" to really talk about subject "B." But most of the audience doesn't understand that, which makes of the theater especially a rather absurd undertaking (if you would make art). Why do I do it? Well, I had my reasons, but I'm not sure I'll do it much longer.

One reason one makes art is to have more control than usual over what goes out and into one. Because if you are making a work of art, you devote a significant number of the available hours of the day to controlling that input and output. So the work of art is always a picture of one's ideal world, a postulated utopia. But again, it is not the "things you are talking about" which constitute the content of that utopia, those "things" are used to talk *really* about something else. So the utopia—there, before your eyes—is unseen by most people. Why both talk about what you are talking about? Ah—to talk about it is to first catch it, so that it can be "displayed" (talked about in theatrical language). To catch it, to make it hold still, you have to kill it. Everything that is talked about (displayed on the stage) is a dead thing. I don't want to "kill" what I REALLY want to talk about (utopia) so I have to talk about OTHER things I don't mind killing and then those dead things are talking about other things that somehow—because they don't have to be "displayed," don't have to be killed.

To EXPRESS something means you first killed it.

One can hardly help generating things that give pleasure. We are built to "take pleasure in." But the effort must be made to try and insure that pleasure will feed not the "I" (which should be the developer of will) but rather the disassociated not-I within us. The not-I is both a more sensitive, subtler, more intricate pleasure-experiencing machine than the I—and the sole field within the person which will NOT degenerate through repeated pleasure stimulation. Clearly the "I," the ego, does so degenerate into sensualism if fed too rich a diet of pleasure. But inside of us there is all that "passes through us" (the other, which is always threatening to disrupt, the selfhood we feverishly hold onto) and that "other" in us takes the pleasure it is fed, breaks it into a hundred small pieces and sends it flying to feed different parts of the energy system which because it is always challenging the coherence of the inner "I," forces us to new efforts of will and invention.

How to feed the "other" in us (the not-I, as opposed to ego) with pleasure? Ahh—but everything in this collection of notes is really speaking to that primary end, dealing with that primary problem.

Ritual as anti-doing, the anti-pole to force (see Erik Gutkind). My life of writing is a ritual, I make nothing through force. I copy certain things (or, let rise certain things) and that doing-so renders me transparent. Erases me as a "force." My work . . . erases me. So, I am not. What I am finally, is a part of the composition that arises through me.

We can postulate two (of many) systems going on inside us.
1) A "receive perception" system (always a clean slate).
2) A memory store.
Those are "imagined" systems, suggesting new ground rules for the game of art-making. As opposed to such an imaginary system there is a more verifiable neurological bi-part system in which
1) Certain neurons PROTECT us against the strength of incoming stimuli.
2) Certain others receive stimuli.
The fact is that we pick up the frayed ends of system 1 on system 2. So I can suggest to myself—write the PROTECTION against noticing, generate gesture of defense against input.
Also:
The perceived may be read in (on) the past (the memory slate upon which past perceptions have left their imprint).
"Pure" perception would go in, and vanish, and be not.
Real perception is resistance to perception.
Can you imagine what kinds of texts, suggested by above procedures, might be generated? Would they not resemble my texts?

Old paradigm: Universe consists of forces that solidify into units (Gestalts, objects, events) to which we *respond*.

New Paradigm: Universe consists of forces that leave traces which are not fully identifiable consciously, of which we see only residual evidence—and if we respond it is an "error" of responding to what we *project* into those traces.

If you believe in 1, your art tries to make something visible, and the life copied by that art is a responding-to-input from the "world."

If you believe 2, your (my) art tries to erase things (because they are obstacles) and the life copied by that art is a "something else" that tries to resonate to inner output.

The TREE of senses.

Man is currently the "seeing" creature—that sense defines him vis-à-vis other creatures, who have more highly developed "lower" senses.

Smell (taste)

Then: hearing

Then: seeing (man's current level)

Then: thinking (the next level, not yet achieved.)

THINKING . . . as a sense. As a way to respond to what is present

. . . by THINKING what is present, rather than smelling, hearing, seeing it.

So, try to make a new art about THINKING—THINKING treated as a *sensing,* as the sixth sense!

Try to imitate (anticipate) the next stage in the evolution of consciousness. What that amounts to is a planned opposition (within the work) or restriction

of organic releases (pleasure): which is also a way consciousness could be thought of—a restriction on immediate release in sensation.

Past achievement of man: to turn "tree" into a sign, which can be held in the head. That's what men have achieved—symbol-making, sign-making ability, in which conscious experience mediates between man and encountered tree.

The next step might be to restrict the emotional release man now gets through his encounter with signs, and so see the sign (object) dissolve into a kind of web-of-association awareness. See the signs become nothing more than polarity-traces. That web-consciousness then mediates between man and signs and he no longer sees the "signs"—just as in encountering the tree in the field he no longer really sees the "tree." Instead of the tree—he flashes the sign in his head, "tree." So in the future—he no longer flashes the sign—but the entire web-of-associations and differences in which tree-sign occurs as an item.

Then: thinking . . . as a sense. In the way that "seeing" now mediates between man and experience—separates him from experience because it translates outside to inside—so thinking could be a similar translator . . . KNOW THAT, and make the ENJOYMENT of that be the art enjoyment. Because to separate himself from nature, and then from experience even . . . seems more and more to be man's destiny! Man *is* the abstracting animal. Keep going.

KEEP GOING!

Program Note for

Blvd de Paris (I've Got the Shakes)

O.K. It's about the rhythmic oscillation, very fast, between insideness and outsideness.

It's about the tapestry (many threads from many sources) weaving itself and reweaving itself. That process. Each moment . . . a unit of joy coming into itself. Things bleed in unexpected ways into other things. A reverberation machine! That's what my plays are! The theme is the "sequence of things," the theme is "everything that wants to be written"—and how everything is secretly present in everything else.

Also, and more hidden (more EMBODIED in the work, like the blood flowing through its veins) there is a certain kind of activity going on in the play—a new kind of activity which postulates thinking as sort of "close to the surface of the body": body mechanisms and manipulations as thinking and perceiving mechanisms and manipulations. A series of manipulations of objects and desires and situations . . . treated by bodies-as-thinking-mechanisms. The plays are about whatever happens when I *AM* in a certain way, functioning on a certain level (which gives me most delight) and I open to you in that delight, my joy wants to amuse you with the fact that things inevitably WILL connect, WILL reverberate with each other. THE WORLD IS A REVERBERATION MACHINE, that's what I show you! I allow it to hum even louder than usual by turning, in my plays, away from mechanical causalness. The theme is to document in the plays a certain kind of "constructed" behavior (my invention) in which mental-acts take place on an outside surface . . . not hidden away inside. Thinking as the product of field-interchange. The joy of making a thing dance in the mind-which-is-outside, in the body and its field!

Unpublished, 1977.

Director's Notes

from Program for *Don Juan*

What is of supreme interest to me in a theatrical text is the manner in which each character is, at each moment, being dissolved and reconstituted by the *languages* which course through him or her. Following the lead of contemporary French theorists, I find it most productive and illuminating to regard the written text of the playwright as the "deposit," the "tracings" of what obsessed him, as an individual, in each instance. These theorists, of course, are Lacan, Foucault, Derrida, Kristeva, et al., who have, I believe, profoundly rethought the problems of what it is that constitutes man, his cultural world and its creations.

A play as written is first and foremost, for me, a language-system. One aspect of the rules of that particular game, when the work in question happens itself to be a "play," is that the language shall, among other things, give shape in our imagination to things known as "characters" and "situations" and "actions." But the language, the discourse, the flow of the text as the outpouring of one obsessed individual (the writer) is the primary aesthetic and dynamic element, I believe. He chooses to pretend he is "talking to himself," as it were, by splitting his language flow into several persons.

And so, the real heart of the aesthetic matter is not, in my opinion, the overall "fable" which shows certain characters in conflict in order to suggest a certain "thematic" content. No.

Might I shockingly suggest that in the world of art, in the dramatic masterpiece (which certainly *Don Juan* is), the heart of the matter is invariably the way in which the playwright (the one who "flows language") allows the expressive thrust of his language to suggest themes,

Unpublished, 1982.

ideas, connections, impasses, and all of the twitches and stirrings of real life and effort? But this is not always neatly parcelled out to appropriate "mouthpieces" who can be taken as a model or analog for certain thematic statements. Rather the thematics embodied in the language of the text pass through and are refracted from character to character so that each character is again "destabilized," as it were. Each character twists and turns under the attack of what "must be said" by the writing flow, and the great play (and I would go so far as to say that this is the very definition of dramatic greatness) shows us "characters" who are as a series of mirrors, reflecting and blending into each other under the impact of the playwright's flow of language.

For that reason, in my production you will see a series of articulated echoes and reflections, as the characters mimic both each other's and their own language and gestures. You will also see Don Juan, as well as others, continuously set against the crowds, so that what is in one sense, at one moment, a monologue, is also then appreciated as an effort to shape a "self" against the background of an always demanding society. This society frequently stands by to enforce the rules of its own language-system, not only of grammar but of laws, behavior codes, and the like.

Now then, the interesting thing about all this is that to think of the dramatic work in this way (as a language-system within which characters are struggling to emerge as "selves") also gives us a model which is appropriate to the social situation of the individual within his society, our own society or Don Juan's. For example, Don Juan is, at moments, a hypocrite. But, in being a hypocrite, he is echoing the hypocritical world about him, serving even as a sort of punitive magnifying glass. He *is,* therefore, greater and more intelligent and more aware than those who surround him . . . even to the extent that he is perhaps "worse."

Although he quite properly isolates and speaks out against the stupidities and false pieties of the society in which he finds himself embroiled, he still is, inevitably so, a product of that society; moreover, the self-defining *language* of that society is unavoidably coursing through his own discourse. He is, therefore, often "falling out of his own positions," as it were. Then, through an effort of reason, he tries to force himself back into his "chosen" intellectual position. So it is with most of the characters: they start to speak, to define themselves, and then the drift of their inherited language invariably takes them to places in which they didn't quite expect to end! Away from themselves! Which is what language does to all of us: we start talking, and the *language* takes over. We end up being ever so slightly (or, sometimes, vastly) untrue to what we thought of as "ourselves" because the available language speaks through us and takes possession of us without our quite realizing it.

I believe that the great playwrights make such a double awareness (of language-system and social-system) the source of their work. And my work as a director is to find continual analogies to the "language-work" of Molière within my own available directorial stage-language (which consists largely of gesture, movement, sound, image, inflection, etc.), so that the theatrical event will become a continual and lively demonstration of the ways in which man is being forced to circulate within the codes and forms and social options which are available to him in a given society. He is victimized by his own (inherited) discourse, trapped by his own language, tortured by his own impossible effort to escape "what his language would have him say!" But the desperate and always relatively unsuccessful human effort is also the exhilarating basis for the life of heightened (if painful) consciousness, which becomes great art.

To see how desperate things are is to understand that a fraction of human effort is never more than partially successful and that human beings are invariably only a fraction of what they might become: such, I believe, to be the bitter insight of all great art. And yet, amazingly, the formal, specifically aesthetic, strategies of art somehow turn the bitter medicine of truth into the energy of great art. This is what, I believe, is often not understood: that it is the "aesthetic," "formal" aspect of art which *really* excites and lifts and is most positive. Such formal strategies which have to do with the ways in which the *art-language* is able to find within the often painful, ugly, bitter materials of its "content" a network of echoes, repetitions, reflections which can be shaped to produce the harmony and counterpoint which speaks to us so that our hearts would flutter with excitement . . . and our minds would dance. Yes! The mind . . . *dancing!* That is the muse I would serve, through the very torturous coursing of this great, bitter, frightening "comedy" by Molière.

REMEMBER—Molière's *Don Juan* was violently attacked as a radical and immoral play when first produced. And indeed, it was radical, it was profoundly upsetting and should remain so today if it is to live as anything more than a museum piece.

REMEMBER—Don Juan lived at the beginning of the Enlightenment, which, to paraphrase Adorno, replaced the religious "myths" of the past with the new "myths" of science and "fact."

REMEMBER—He revolted against a society, both oppressive and one-sided, being himself a harbinger of the very Enlightenment (the rule of reason and fact: "two plus two equals four" as the ultimate belief) that seems to smother us today under the technology, mass media, and conglomerate corporate structures which prove to be the Enlightenment's unavoidable progeny.

REMEMBER—Don Juan finds no way to effectively counteract repressive and constricting religious dogmatism and its accompanying hypocrisy. He is therefore forced to merely "act out," "externalize" the image of his rage and revolt and translate his rejection of the world about him into nothing more than the *show* of rebellion. In doing so, he effectively only tears himself apart with his own anguished justifications of libertinism and rationality, he destroys himself—victim both of his own limitations and of a society that doesn't want the boat rocked.

REMEMBER—This play, written by the King's favorite, created such a storm of protest after its initial engagement that it was never again performed during Molière's lifetime. Certainly Molière knew what he was doing when he wrote this very angry and disturbing comedy. And certainly it is our task to give birth again, for our times and audiences, to the searing agitation and anguish that produced this very important, very austere, and very radical "comedy."

Ages of the Avant-Garde

Was I ever avant-garde? I suppose so, since when I began I did consciously reject everything around me called theater, and tried to build my own counter-world of theater. Having now built it, after twenty-five years, I suppose I now have a "world of theater" in which I live (my own), and as much as I'd like, in principle, to escape and upset that apple-cart also, it is of course very hard to do while yet remaining true to one's particular vision. (See Otto Rank's discussions of the disruptive artist who, having attained a modicum of success with work that at first "attacked the world," then finds his subject to be his ambivalence and struggle with his own now "acceptable" work.)

So some others still think of me as "avant-garde," while others think of me as a now fairly conventional "modernist." But I believe the true fact is that as I grow older (now fifty-five) I have "aged" into a more open confrontation with what has always been the true, half-hidden subject of my work from the beginning—the confrontation with death (absence, hiddenness, emptiness—the "pretense" of life in the face of all this—the theater as the manifestation of this "pretense" in the face of death).

I believe that the serious resistance my work has generated (without wishing to minimize my appreciation for those who do and have supported it) has to do with a sensing of the way my work has always been very aggressive in maintaining that life as we know it (and as normal theater knows and presents it) is an absolutely silly, childish (and understandable) avoidance of the emptiness at its center. So I believe that my work originally seemed "avant-garde" because it aggressively mounted a disassociated consciousness trying to "echo" the manner in

Originally published in *Performing Arts Journal* 46 (January 1994): 15–18.

which the "center does not hold"—ever—as all our so-called coheren-
cies are mere wobbles of artifice around a central black hole. And as I
grow older, this central awareness tries to surface in my work in more
naked form—pulling me back, I suppose, into modernism—as opposed
to a postmodernism that "dances the dance over the abyss" without—
as I see it—paying homage to that abyss itself, the terror of which generates all our constructions and inventiveness.

At fifty-five, since energy is less than at thirty-two (I was a late be-
ginner, my first anything was only mounted at thirty-two!), I think one
tries to become "economical"—psychically as well as physically. I can
no longer imagine "killing myself" as I used to with the delight of
mounting a huge production that demanded an "impossible" amount
of effort from me. But I don't see this as a loss, rather as an appropriate
attempt to cut through the frenzy of glitz and excitement to get to the
delicate seed of things. I know some people lament the fact that my
work has been getting simpler and less flamboyantly theatrical. But for
me—it's a new kind of strength. I feel I have touched something deep
that no longer needs the support of all the fireworks I used to provide.
I'd like to think that makes me less avant-garde—but MORE radical
and unrelenting. In other words, as aggressive as ever—but perhaps
more secretive. Should the "secret" replace the "avant-garde"? Is that the
only way to avoid being co-opted? I think history suggests such may
well be the case.

To be secret means to accept the marginality of one's position in
American culture. To finally accept that one's work will never speak to
great numbers of people—and to find ways to intensify the work (en-
courage it to implode upon itself, rather than to explode and touch
places throughout the whole culture—my dream at twenty).

Of course, the belief persists (hard to wipe out) that an intensely
CONCENTRATED work will have deep, if not wide, effect in the
long run. And again—there is evoked in my mind a relationship between that "deep" of the deep effected of concentrated intensity, and
the "deep" that is the unfathomable deepness of death, which kind of
"wipes out" all other considerations. I believe absolutely that as I get
older, I get closer to the truth. How can I help it? Death becomes more
"unavoidable"—less "forgettable." That being the case—do I automatically become MORE avant-garde? I think so, in the following sense.

To my view, the avant-garde classically rejected the forms of art
which existed because those forms were experienced as a kind of lie, a
kind of denial or suppression of the basic energies driving mankind. If

you accept, as I do, that the basic driving energy is fear (which fully explored reveals the fear of you-know-what behind all others)—then the avant-garde has always been an attempt to allow the disruptiveness of death to surface within the patterns of that "normal" art that tried to pretend it has defeated time and death with the eternalness of beauty, harmony, and form. (Even in tragedy, the "form" conquers all.)

Getting old, within the context of the avant-garde, means therefore getting closer to the subject one originally dared to broach by trying to splinter the forms of inherited art. But now—old—one is on the inside (almost) of that subject. There is nothing left to break, the walls (for oneself—not for those still young) have indeed tumbled. And inside the tumbled down walls one discovers—guess what—flowers. I don't fear senility. (I find its approach hinted at, as I occasionally forget names, as I find that only in the early hours of the day is my mind able to really "ignite" as I write my writing.) I don't fear senility, because I sense it as a kind of final "pruning" that allows a final, rarest blossom to flower. I who in my youth HATED flowers.

But I've been very lucky. For twenty-five years I've been able to pursue a certain something that I wanted to see happen on stage. Whether I too, in order to continue, have been to any extent "co-opted" by my audiences or the need to get grants or anything else—that's for others to say. But it seems to be that getting older in the context of a culture that is hostile to the avant-garde, but decidedly NOT hostile to youth, means that the avant-garde in America is often confused with youth—hence the current assertions that various manifestations of pop-drug-edge culture have absorbed the avant-garde and / or made it irrelevant.

In fact, the "official" avant-garde movements of the past, from Surrealism to Dada to Futurism to Expressionism to Minimalism, if understood in depth, with very few exceptions (excluding Pop Art, for instance) have been movements that tried to give material body to the spirit that moves (and therefore disrupts) man. So the avant-garde of course still exists, and one disappears out of society, and into "it," when one ages—with one's eyes open wide.

I should add as postscript two practical matters. (1) The avant-garde has always used dedicated young people as "cannon fodder" in the war against the establishment. Actors, technicians, etc., have to be young (with rare exceptions) to work on the edge. Therefore, as I grow older and my texts express the concerns of an "older" consciousness, I find it harder and harder to find strong performers with a maturity equal to my own who are willing to work for the low pay and relatively small audiences, etc. For me this is the greatest disadvantage of aging in the avant-garde. (2) In the past, in order to lend a certain veneer of "respectability" to my "esoteric" work, I accepted a certain number of jobs directing plays for major producers or institutions (like Papp, Brustein, etc.). And it worked—in the sense that it made me well known enough that my more "difficult" work had to be taken a bit more seriously. But now, as I get older, and don't have the "need" to do the classical work that never really interested me as much as my own—I have the feeling that people tend to wonder: whatever happened to Foreman? He's still doing those "little" plays for his OWN theater after all these years? He HASN'T entered the mainstream? Well, this country doesn't have a continuing tradition of the avant-garde—it has only, as I've said, a tradition of "Hey! What's new?" so I still feel the pull to occasionally do a mainstream work—with the feeling that while it may still help get some people to look at my Ontological-Hysteric work, it nevertheless pulls me away from that center where I should now be spending my remaining time, deepening my work as best I can.

Samuel II

Samuel was a writer who knew he had nothing to say. And knew also (did an angel whisper it over his shoulder) that only by plunging directly into that fact would he discover his strength.

Nothing to say—The air hovered in front of (outside) his window and vanished, like air does. His feeling for other people was like glass, like Marie was glass and could be touched, but the invisibility of what he was touching barked back.

Marie had nothing to say to him, and so it was like an echo, though it was like a noise, and Samuel accepted totally (that was his effort or his adventure) everything momentarily visible on the other side of that dumbness (a sheet of glass) that seemed to separate him from everything he could see the minute he touched it (like glass).

. . . . .

Samuel had nothing to say, nothing to give to the world.

The left side of things entered him and passed through like butter, pulling the lips after it like waves of the ocean, and Samuel felt himself licking the ideas that went too fast to catch.

So the tongue—de-throned. So the language, de-mounted and secret, speeding through the streets, convertible style, and Samuel's weary arms ached, holding out into space, hands that wanted to seize something.

This version unpublished, 1994. Later expanded and published in *No-Body* (Woodstock, N.Y.: Overlook Press, 1997), 163–206.

And those same muscles, gripping the pencil that fell from his fingers onto the page, left nothing but the tiniest evidence.

Samuel smiled to himself in the mirror. Where was that? Quick, it flashed by as the ocean flashed by, shaped like a wave, no shape, just something that undulated, stealing from Samuel, theft as reward.

. . . . .

Samuel, floating on the stream of events, had nothing to say about the events. His head exploded (oh that was his wish) but no head exploded, indeed, events wrapped seemingly tighter and tighter around his cranium.

He lifted himself from sloth and depression one way only. He disappeared into the "before I was born and after I have been dead," but he was not very good about such lifting, it was only a dream.

Oh, the escape into a dream. But, he explained to Estelle, not an escape, a slow and meticulous pushing back of a heavy door. That took effort (great effort) and the results were small (so small) and the two—efforts and results—factored together, made Samuel feel himself a hero of sorts, but he was aware of the dangers of vanity and for that reason called, internally even, no attention to himself.

Samuel had nothing to say, his heroism, self-judgment from a position of rigor.

. . . . .

Samuel is asked to come into the midnight arena of self-regulation.

The double clocks doubt, that is their best, beloved function.

Arenas of stairs, starting in direction.

Choose self for a narrative fiction.

Oh, none of this possible.

Therefore

Into the flesh:

Where the word leads the adventurous choice,

Who is without honor.

Down valleys

Down through open windows

Home-let

Where problems lie.

The grain most alert

self focal dissolve

sea on sea

We-weave

of the aberrant, too-much

Almost.

the final construct

Almost:

The name itself:

Almost.

. . . . .

Samuel had nothing to say to the telephone, but it rang, and he spoke through it.

A voice rang like an echo. Samuel chewed on it, and somebody else said "speech."

The internal dictionary was pulled forward, up through the bone itself, confusing mouth, nostrils—was this speech (?) or merely breath and its discomforts. Breath, with its promise of easiness confounded.

Samuel re-plunged into the re-played aria of suggestible selves and, lost finally, in that network, might as well stop talking which goes on, anyway.

. . . . .

Samuel was a writer who had nothing to say. That was his strength.

A heavy stone was rolling down from a hill, near, and soon would be on top of him totally. That total, that thing, rolling over him, that was both his opportunity and his bad accident.

Fly, Samuel, fly to your destiny. Escape not, except that the escape not is an escape from escape—up to now—having things to say in a pretend way, even under a pretend name which is Samuel even if it is the real name, Samuel.

Samuel had nothing to say, his strength, which returned him to his own name as being a beginning and ending of something he could be involved in completely.

Samuel, Samuel, he self-said. There were many places and ways he could inscribe that name that was his.

Samuel sinking into Samuel, which was the case of a name that was entered, was a creating of space.

. . . . .

In front of the white curtain, strung up to answer Samuel's own anticipation of space.

. . . . .

He knew that the babble of words babbled something quite beyond itself. And yet. And yet.

Silence also pushed through him and searched for the light. Silence plowed through his tongue, his eyes and nostrils, looking to lie exposed to external light, as if an inner light desired full union with that other light.

So babble rose, from a certain source, sped into space but a true bubbling that bubble-like, babbled into the shape that was explosion after explosion releasing, centering the call to silence with each sphere of punctuated, something-to-be-said, was said. And those pin-pricks of a million calls to silence did call forth the silence from a depth that either was or was not the same depth.

So silence and babble, so both, so linked, so opposed and coordinated, in that new, next attached level of babble, laid forth blanket-like under the light, always the light re-showing itself outside which was only a way, finally, of being inside.

My Head Was a Sledgehammer

A large room with pink walls, a kind of classroom perhaps, with many blackboards on the walls. But there are also eccentrically painted gray shapes, suggesting large ink-blots evoking memories of animals or strange geographical continents, and black horizontal lines of varying thickness, which give an added Egyptian-like texture to the walls. Here and there tall bookcases reach to the ceiling. Roses creep up the walls and around corners. and tiny flags of no specific nature are hung here and there. At the rear of the stage is a long railing, a foot downstage of the rear wall, much like an exercise bar in a dance studio. From the ceiling hangs a double-faced head, which revolves slowly throughout the play. There are some small carpets, defining areas of action, and some thin red poles with lights on top. It is a lecture room, but it is also a laboratory, or a studio, or a gymnasium—all at once.

The play begins as piano music repeating a short child-like phrase is heard over what sounds like chattering crickets. Everyone who speaks wears a radio microphone on a headset, enabling them to speak softly and intimately throughout the play. The music—repeating loops made from brief musical phrases—continues through all moments of the play, ever changing, ever returning, giving musical underscoring to all the dialogue and occasionally rising in volume to punctuate a dance or a complex manipulation of objects.

Four gnomelike creatures with red hair, tall caps, beards, and eyeglasses appear on tiptoe. They creep onto the stage and go to different points in the room, from which they each pull a white dotted string from retractable

Originally published in *Theater* (Yale) 25, (spring / summer 1994): 87–101. Reprinted in Richard Foreman, *My Head Was a Sledgehammer: Six Plays* (Woodstock, N.Y.: Overlook Press, 1995).

spring devices that keep all the strings taut. At the end of each string is a small rubber ball, and the gnomes crouch down and dangle the balls over seemingly random spots on the floor. At the same time, the FEMALE STU-DENT *enters, a young woman with long dark hair, ankle-high socks, and a beret. She sets down a briefcase and puts on a pair of white gloves. At the same time, the* PROFESSOR *enters. He is a burly man in a soiled white shirt with the sleeves rolled up past the elbows. He wears a thick, studded belt. On his head he wears a tight-fitting white bandanna, and over what look like motorcycle pants he wears athletic knee pads. His face is covered with many small scratches, and he seems to resemble a member of a motorcycle gang as much as a distinguished professor. He stops as he enters the room, seemingly hypnotized by a single white glove that lies carefully spread out on a white cloth that has been laid on the floor directly in his path.*

FEMALE STUDENT: Okay, Mr. Professor—

With a tiny yelp of pleasure, the GNOMES *all fall to the floor, still holding their strings. The* PROFESSOR *turns to study them. Then he circles back to the glove, bends down, and puts it on.*

Are you as weird as I think you are?

PROFESSOR: *(Turning slowly to confront her)* Not likely.

The music rises, accompanied by a voice chattering rapid gibberish. The MALE STUDENT *runs on suddenly and the* PROFESSOR *whirls to face him. The* MALE STUDENT *is dressed in knickers, carries a briefcase, and on his head sports an army fatigue cap. The* PROFESSOR *runs from him to the* FEMALE STUDENT, *who stamps down on his foot, and the* PROFES-SOR *yowls in pain and runs to one of the blackboards and begins to erase—though there is nothing to erase—and the* MALE STUDENT *follows him to the blackboard to observe, whereupon the* PROFESSOR *drops his eraser and slams himself with his back against the blackboard to watch for a further attack. During this the* GNOMES *have retracted their strings, and now all squat at the rear wall, holding the exercise bar, and the music is quiet again.*

FEMALE STUDENT: Disturbed? You hear voices, Professor?

PROFESSOR: I'm lucky that way.

He takes a few steps, and she shadows his move.

FEMALE STUDENT: I'd say, weird.

PROFESSOR: I feel connected.

FEMALE STUDENT: Connected to what?

PROFESSOR: *(As both students stalk him)* To the source of my voices.

FEMALE STUDENT: Okay. That's weird.

PROFESSOR: . . . Are you weird, or am I weird?

FEMALE STUDENT: What's the criteria?

Three of the GNOMES *run forward, each carrying a tray with a small printed card. The* PROFESSOR *and the* STUDENTS *each pick up a card, glance at it, and toss it to the floor, whereupon the* GNOMES *retrieve the cards and go back to the exercise bar.*

PROFESSOR: *(Addressing the female student)* Let me invite you— *(Takes an apple from a bookcase and holds it out to her)* come a little closer.

FEMALE STUDENT: In what sense, Professor? *(She runs to him carrying a rope, and quickly entangles him with one hand behind his back.)*

MALE STUDENT: The question is: what can he do with one hand tied behind his back?

The PROFESSOR *unwinds himself as the* GNOMES *exit, and he proceeds to erase the already erased blackboard, as the* FEMALE STUDENT *goes and lies down on one of the tiny carpets. The* PROFESSOR *stops, and as the* MALE STUDENT *goes down on all fours to crawl toward the* FEMALE STUDENT, *the* PROFESSOR *also sinks to the floor.*

PROFESSOR: My secret desire is—win love from many beautiful women.

FEMALE STUDENT: *(Propped up on an elbow to look at him)* I'll pretend I didn't hear that sentiment, Professor.

PROFESSOR: I'm speaking of things closest to my heart, madam.

The four GNOMES *run in with long silver wands and take positions around the* FEMALE STUDENT, *using the wands to point at a watch on her wrist.*

MALE STUDENT: Notice what's on her wrist, Professor?

In slow motion, the GNOMES *fall to the floor.*

PROFESSOR: A very beautiful watch.

As soft organ music begins and GNOMES *retire.*

MALE STUDENT: Hey—who do YOU pray to, Professor?

PROFESSOR: I don't.

MALE STUDENT: What kind of religious ceremony?

PROFESSOR: I don't have any religious observations.

MALE STUDENT: God plays no part in your life?

PROFESSOR: None at all.

MALE STUDENT: There is no God in your life, Professor?

PROFESSOR: None at all.

Suddenly one of the GNOMES *rolls a cabinet into the room at great speed. It comes to an abrupt stop. Displayed in the cabinet is a violin.*

FEMALE STUDENT: What a beautiful violin, Professor.

MALE STUDENT: *(As if experiencing a vision)* My God—

FEMALE STUDENT: Do you play it?

PROFESSOR: I scratch out noise.

FEMALE STUDENT: I'll bet you play better than that, Professor.

PROFESSOR: I scratch out terrible noises, which I like doing in private.

FEMALE STUDENT: You're saying that just to provoke me, Professor.

PROFESSOR: Not at all.

An electronic tone rises, and the PROFESSOR and the MALE STUDENT put hands to their foreheads, moan, and fall to the floor, as the FEMALE STUDENT crosses and pulls the violin from the cabinet. She studies it in her hands, as the PROFESSOR and MALE STUDENT recover and run from the room. Just at that moment, the GNOMES run in and snatch the violin from the FEMALE STUDENT and run out of the room with it. She chases after them, and the PROFESSOR and MALE STUDENT reappear and join the chase. Then, alone together, the three embrace one another despondently. Each kisses his own right hand, and three GNOMES sneak back into the room and go to pull the strings from the wall. The PROFESSOR and his STUDENTS have been caressing their own faces with their kissed hands. But when they see the GNOMES dangling the rubber balls at the end of the strings, each goes slowly on tiptoe to one of the balls, to take a position leaning ecstatically against a wall with a ball now placed close to each of their bodies. The tone is very loud, and the three of them stare up into space, pointing one finger to their foreheads.

FEMALE STUDENT: *(As the tone softens)* What does your internal time say now, Professor?

PROFESSOR: It doesn't speak to me, it manipulates me in other ways.

FEMALE STUDENT: Okay, Professor.

They all leave their positions.

Imagine a completely different play called . . . *(Tries to think, then comes up with a title)* "Fingers Alert." What could happen in such a play?

The tone rises again, and while the two men cover their ears, the FEMALE STUDENT runs to the blackboard and runs her fingernails down its sur-

face. Then everyone runs to new positions as the music changes. The FE-
MALE STUDENT *has her back to the room. The two men sneak toward
her and one softly whispers "Cookoo!," which makes her whirl in irritation.
She picks up a metal dog leash from the floor, bending over provocatively,
then turning to threaten the men, who are muttering happily to themselves—"I guess we showed her a thing or two."*

FEMALE STUDENT: Imagine a play called "Dogs on Duty."

*The two men take a step back and growl, and then reach for books high on
the bookcase, but unable to control the books, a few tumble to the floor, as
the* FEMALE STUDENT *is gliding about the room, trailing a dog leash on
the floor behind her provocatively.*

What could happen in such a play?

Disco music rises as she circles the room. The GNOMES *arrive with plates
carrying cards that are again read and tossed away. One* GNOME *places a
chair wrapped in brown paper on the little carpet, and then the* GNOMES
go to the rear and rest on the exercise bar. The PROFESSOR *has acquired
another apple, which he holds out toward the* FEMALE STUDENT.

PROFESSOR: Did you ever see this in real life, madam?

FEMALE STUDENT: *(Approaches it slowly to study it)* An apple!

She and the GNOMES *and the* MALE STUDENT *form a line, and they all
cover their eyes.*

PROFESSOR: *(Re-presenting the apple)* Remember? Did you ever see this
in real life?

Everyone in line slowly looks toward the MALE STUDENT, *who covers his
private parts with his hands, and all but he and the* FEMALE STUDENT
slowly fall to the floor as the childlike piano tune returns. Then the FE-
MALE STUDENT *grabs the chair wrapped in paper and pulls it noisily
from the room as the* GNOMES *rise and run out after her. The* MALE
STUDENT *comes down to watch the* PROFESSOR *erase the already erased
blackboard. Then he begins erasing one right next to it. The* PROFESSOR
is jealous and pushes him away and starts erasing on this other blackboard.

The MALE STUDENT *tries again, and as the* PROFESSOR *reinvades his blackboard, he gives up. The* PROFESSOR *stops and studies the* MALE STUDENT.

PROFESSOR: You just lied to me.

MALE STUDENT: I don't lie, I don't tell the truth. I'm here to do neither of those two stupid things.

PROFESSOR: What are you here for, sir?

MALE STUDENT: To be slippery on things, Professor. *(He twists his legs and mockingly pretends to slip, falling to the ground with a shriek that turns into a laugh as the* PROFESSOR *runs across the room, opens a briefcase, and pulls out a stack of papers.)*

PROFESSOR: Well, here's what I'm here for.

He starts throwing individual pieces of paper into the air, and as they flutter to the floor, the GNOMES *run in and collect them and run to the walls to hold the sheets of paper high up on the walls.*

I want to be in a place
from which truth—
(He whirls ecstatically and runs across the room, only to have an accident—he bumps into the wall and cries out in pain as he holds his damaged nose.) Ow! Now we have a problem. Truth: gushes forth.

The MALE STUDENT *barks twice, breaking the mood as the* GNOMES *turn from the wall to watch.*

I want to be a place THROUGH which truth—

The GNOMES *run from the room, clutching their papers as the* MALE STUDENT *barks again.*

I want to be a place through which truth passes.

He has come and captured the MALE STUDENT *in a headlock, but the student immediately escapes.*

But when it passes through ME—
stripping off its protective cloak—
Ah, but why do I wear my protective cloak?
Because truth revealed, believe it or not, takes the unfortunate shape
of everything that isn't true.

He crosses to the MALE STUDENT, *who is now erasing one of the blackboards already erased.*

This happens! THIS REALLY HAPPENS!

As the MALE STUDENT *turns to face him, the* PROFESSOR *turns away sadly.*

But if mere actors speak this, then it no longer happens.

MALE STUDENT: *(Inspired to imitate his professor)* Me too. I want to be, a place through which truth passes. *(Goes to take down a book.)* Did I get that right, Professor? *(He bounces gently across the room, singing a scale to himself. Then turns a page and finds the passage he was looking for and reads.)* "I want to be a place through which . . . truth passes."

The FEMALE STUDENT *is crossing behind them, studying an apple she holds in her hand.*

PROFESSOR: Turn the page.

MALE STUDENT: *(Ecstatically hugging the book to his chest)* I'd like to linger a bit longer over this very particular page.

PROFESSOR: Turn the page.

MALE STUDENT: *(As he does so)* What does it say on this page, Professor?

PROFESSOR: Wait a minute— *(He crosses to the* MALE STUDENT, *momentarily distracted by a big pink disk that rolls in behind him, manipulated from behind by a* GNOME. *Then he recovers his composure, and goes to whisper to the student)* That's my line. *(Grabs the book away and reads)* "What does it say on this page?"

MALE STUDENT: "Eat me"?

PROFESSOR: Please?

He checks the book, sees no such line, and hands the book back to the MALE
STUDENT, *who proceeds to tear a page out of the book and, with a flour-
ish, eats it.*

By the way, Professor—

The FEMALE STUDENT *comes rushing forward to present the apple to the*
PROFESSOR, *who scurries out of the way and runs to fall back ecstatically
against a wall, as a* GNOME *pulls a string from the wall to hold on the*
PROFESSOR's *knee, and the* MALE STUDENT *retreats to pose spread-
eagled against the pink disk at the rear of the room, continuing all the
while to chew the page torn from the book.*

—do *I* consider you a place through which truth passes? You'll never
know, Professor, because you can't get inside my head. *(Turns away)*
Let's make a guess—

FEMALE STUDENT: *(Running forward)* What's a guess?

PROFESSOR: Somebody should write this down so I don't forget. *(He
whirls away to find a piece of paper, but miscalculates and bangs against
the wall.)* Jesus Christ—! *(Holding his head, he realizes the collision
was very stimulating.)*—this is the most interesting thought I've had
in a long time. What's a guess?

MALE STUDENT: *(Immediately copying his* PROFESSOR, *he comes and
bangs against the same wall)* Jesus Christ—! *(Spits out the paper he has
been chewing and holds his stomach)*—this is the most interesting meal
I've had in a long time. What's a guess?

PROFESSOR: Do you think I can make this lady disappear? *(He gives
the disk a shove and it rolls out of the room.)*

FEMALE STUDENT: I do hope somebody gets hurt.

PROFESSOR: Why?

MALE STUDENT: Hey, do you know the name of every book in your oh-so-extensive library, Professor?

PROFESSOR: I believe I do.

FEMALE STUDENT: I doubt it.

MALE STUDENT: Close your eyes, Professor. Then feel your way to the bookcase. Take out a book without looking.

In trying to do so, the PROFESSOR *again bangs into a wall. But as he recovers from the collision, the* GNOMES *arrive to offer a white cloth to the* STUDENTS *and the* PROFESSOR. *They each take one and hold it in front of their eyes as they take three tentative steps forward. Then as the music rises, the* GNOMES *steal the cloths and run from the room, as the* PROFESSOR *and the* STUDENTS *spin dizzily. As they spin, the rolling cabinet has reentered.*

MALE STUDENT: As you can clearly see—

PROFESSOR: Where?

MALE STUDENT: *(Holding his forehead as the* FEMALE STUDENT *takes an envelope from the cabinet)* I already wrote the name of the book I predicted you'd pick, and sealed my guess in a white envelope.

FEMALE STUDENT: *(Opening the envelope and reading)* It's titled— *(Looks up at the* PROFESSOR*)*—you're not going to believe this.

MALE STUDENT: Oh, I believe it.

FEMALE STUDENT: "Dossier of Fear."

PROFESSOR: I have no such book in my entire library.

MALE STUDENT: Hey, look again.

He collapses to the floor as the PROFESSOR *runs to the cabinet, and the* FEMALE STUDENT *tears the envelope to pieces and throws the pieces into the air.*

PROFESSOR: *(Looking at a book he has taken from the cabinet)* You're quite right. The book is entitled "Dossier of Fear." But I'm convinced that heretofore I had no such book. *(Leafs through it)* Its pages are all blank pages.

A GNOME *snatches the book from the* PROFESSOR *and runs out of the room.*

MALE STUDENT: I could have secretly hidden that book in your collection. Isn't that possible, Professor?

PROFESSOR: Would you do that?

MALE STUDENT: Well, imagine a play entitled "Broken Promises."

The GNOMES *run in with red apples on dinner plates and place them on the floor, as choral music rises. The* PROFESSOR *and* STUDENTS *each take a rubber mat from a shelf, spread it on the floor, and vigorously wipe their feet on it, grunting in accompaniment. Then they put away the mats, and the* PROFESSOR *seizes a large shapeless object wrapped in brown paper, which he holds out toward the* FEMALE STUDENT, *who has run to the exercise bar, where she swings her lower leg in a circular propeller motion. The* MALE STUDENT *grabs the object from the* PROFESSOR *and hides it in the bookcase as the* PROFESSOR *runs to observe the* FEMALE STUDENT *from another angle. She immediately stops.*

MALE STUDENT: Welcome. Let me tell you a story in the center of which I hide a very personal message.

FEMALE STUDENT: Excuse me but—how will we be able to recognize such a private message?

MALE STUDENT: Well . . . *(Reaches out toward the plates with apples, and the two others do likewise.)* Imagine a play—

As the music rises, all three grunt and lurch in place toward the apples. The GNOMES *run in to grab the plates and apples, and run out of the room banging the plates together, making a noise that causes the* PROFESSOR *and* STUDENTS *to hold their ears and howl in pain. They race to the bookcases, and each grabs two plates with handles on the bottom, enabling them to be held like orchestra cymbals. The music stops, and they awkwardly hold the cymbal plates, looking at one another. Then they slowly hide them behind their backs. The* FEMALE STUDENT, *especially embarrassed, runs from the room, and the* PROFESSOR *takes a few steps after her, to watch her exit. The* MALE STUDENT *has begun to erase an already erased blackboard, and the* PROFESSOR *turns to address him.*

PROFESSOR: There might be enough time here to explain something.

MALE STUDENT: What?

PROFESSOR: Did you ever heretofore explode, Professor?

MALE STUDENT: Is this called talking to myself?

They retake the cymbal plates, and as choral music rises, each runs to a wall, smacks the plates high up against the wall, and holding them against the wall, revolve their hips three times, grunting in rhythm. Then they whirl to face each other.

PROFESSOR: Try using my real name.

MALE STUDENT: I don't think names are a real issue, vis-à-vis a man who wants to genuinely explode.

PROFESSOR: On the verge?

MALE STUDENT: Hey, why not take me up on that?

The childlike piano tune returns, as a GNOME *rolls in a wooden cart in which the* FEMALE STUDENT *is riding, as another* GNOME *wheels in a white folding hospital screen, which the* PROFESSOR *studies for a moment, then folds about himself so he is hidden inside. By this time the* MALE STUDENT *is bent over holding onto the railing of the cart, and the* FEMALE STUDENT *hits him three times with a riding crop. He cries out in pain and runs away.*

PROFESSOR: *(Inside the screen, softly)* "When stones are shoes, the bottoms of the feet go deaf."

MALE STUDENT: Do I have your permission, at least, to get torn to pieces by contradictory forces?

PROFESSOR: *(Opening the screen)* Of course you have my permission.

The FEMALE STUDENT *reenters with a large stuffed horse, which she transfers to the* MALE STUDENT, *who lets it fall to the ground at his feet. A* VOICE *is heard on tape, repeating the phrase "Remarkable people never depend on self-revelation." The repeated phrase is joined by organ music as the* MALE STUDENT *picks up the horse and lifts it over his shoulders, its legs hanging down along each side of his neck. At the same time, the* PROFESSOR *places an apple at the* MALE STUDENT'*s feet, and the* PROFESSOR *and the* GNOMES *stand in a semicircle around the apple, pointing to it with silver wands.*

MALE STUDENT: I was hoping you'd offer me a handkerchief.

PROFESSOR: Why of all things a handkerchief?

MALE STUDENT: Can't you guess?

He slowly revolves in place, as the PROFESSOR *and the* GNOMES *slowly fall to the floor, with wands still pointing toward the apple.*

A handkerchief to stop the bleeding.

PROFESSOR: *(From the floor)* From which wound, pal?

MALE STUDENT: Ah, some of them can't be reached. So I'll have to try this one conveniently here in the palm of my hand.

FEMALE STUDENT: Imagine a play called "The Pretend Hat." *(Takes a man's felt hat from a bookcase and places it on her head)* What could happen in such a play?

PROFESSOR: That's a very, very old wound, madam.

MALE STUDENT: I was hoping you'd offer me a coat.

PROFESSOR: Ah, hearts of ice.

MALE STUDENT: *(Smiling, he looks upward, as lively music begins to filter into the room.)* Oh, Professor, you remembered.

PROFESSOR: *(Hearing the music, his hands travel up as if he were going to perform an exotic belly dance. His hips sway with the beat.)* I remember nothing, pal, I just . . . *(The music is loud now, and he has to shout)* I GO, with the FLOW—

To very aggressive, marchlike jazz music, PROFESSOR *and* MALE STUDENT *do a strange dance where they grunt to the beat, pumping their arms like suction pumps, then strutting around the stage doing an exaggerated breast stroke, then standing in place again, grunting and working their arms. Perhaps it's the memory of a bizarre military career. The* GNOMES *have appeared in the entrance to cheer them on, but a moment later the* GNOMES *are racing for the horse, which was dropped as the dance began. They lift it high in the air and run with it to one of the blackboards. As they rub its nose into the blackboard, both the* PROFESSOR *and the* MALE STUDENT *rush to the same blackboard and begin erasing. The* GNOMES *drop the horse and pull it by its legs to the other side of the room. In the rear, the* FEMALE STUDENT *has reentered pulling the cart, which now has a large wooden chair loaded in it upside down. She comes to a stop and watches. She is wearing a graduation robe and cap with tassel. The* GNOMES *run from the room, and the music turns quiet and threatening. The two at the blackboard sense her presence and stop erasing, but they don't want to look at her.*

FEMALE STUDENT: *(Smiling, looking straight ahead)* Thank you, Professor.

PROFESSOR: For what?

He and the MALE STUDENT *go back to erasing.*

FEMALE STUDENT: I'm thrilled to be included.

PROFESSOR AND MALE STUDENT: *(They stop and look at each other.)* In what?

FEMALE STUDENT: By the way, Professor Number Four . . . ?

She holds out a set of car keys and rattles them. The PROFESSOR *and the* MALE STUDENT *race to separate bookcases, pour themselves a shot of liquor, and toss it down in one gulp.*

PROFESSOR: *(Gasping for breath after the drink)* I needed that.

MALE STUDENT: *(Gasping)* I needed that too.

FEMALE STUDENT:—is still parking the car.

PROFESSOR: Parking the car?

MALE STUDENT: Then we could get rid of the horse.

FEMALE STUDENT: What horse?

PROFESSOR: Better not.

MALE STUDENT: Why not?

They both take another stiff shot. Then the PROFESSOR *comes up behind the* FEMALE STUDENT *and smiles sweetly.*

PROFESSOR: Have a seat. *(He looks about quickly, and is embarrassed to see there is no chair, so runs back for another shot.)*

FEMALE STUDENT: Hello again, Professor.

PROFESSOR: Where's Professor Number Four?

MALE STUDENT: That's been covered.

PROFESSOR: With what?

FEMALE STUDENT: Parking the car, Professor.

PROFESSOR: Parking the car.

MALE STUDENT: *(To himself, amazed)* I knew it, I knew it—

PROFESSOR: *(Racing to the female student)* Have a seat.

FEMALE STUDENT: *(Hesitating, with just a hint of seductiveness, she begins taking off her cap and gown)* I already did.

A tone rises as the PROFESSOR *and the* MALE STUDENT *slowly fall to the floor muttering to themselves, "Oh, yeah," overcome by her partial disrobing. A* GNOME *comes to take the chair from the cart, setting it on one of the small carpets, as the* PROFESSOR *struggles to his feet, holding his head.*

PROFESSOR: It must be that something powerful happened to me. I heard this very desirable woman say out loud to me, "I already did."

FEMALE STUDENT: But I already do.

PROFESSOR: Vis-à-vis, you know—

MALE STUDENT: Horses!

PROFESSOR: No. Chairs.

Music is rising that evokes the image of many GNOMES *busily scurrying through the city.*

I wanted those words to be magically invested with something powerful enough to throw me to the floor.

FEMALE STUDENT: I did.

PROFESSOR: I tried to make it happen.

FEMALE STUDENT: *(She approaches the* PROFESSOR*)* I did.

She hits the PROFESSOR *and he falls to the floor. The* GNOMES *appear and accompany the fall with a "Wheee!" of excitement. Then she approaches the* MALE STUDENT.

I did.

She hits him, and he also falls, accompanied by a "Whee!" from the
GNOMES. *Then the* GNOMES *rush forward and help the two on the floor
to rise, and they do a whirling dance with the* PROFESSOR *and his* STU-
DENTS. *The* PROFESSOR *shouts out over the dance—*

PROFESSOR: I remember I said, "Where's Professor Number Four?"
And you said, "Parking the car, parking the car, parking the car—
which might take a very long time!"

The dance music stops and the GNOMES *run from the room.*

MALE STUDENT: Hey, how long?

FEMALE STUDENT: *(After a pause)* A very long time.

PROFESSOR: What's the estimate of a very long time? *(Ominous music
begins softly, and everybody finds a new corner of the room in which to
feel more secure.)*

FEMALE STUDENT: Let's just say—time doesn't exist for me.

As the cabinet rolls into the room, the MALE STUDENT *puts on a dunce
cap, and the* PROFESSOR *takes a rolled-up chart out of the cabinet, waits
for a moment, then lets it unroll dramatically to show a brightly colored selection of automobiles painted on the chart. There is a pause while the two*
STUDENTS *try to make sense of it.*

MALE STUDENT: Okay. Something about a horse—

PROFESSOR: Wrong. *(Then he thinks)* But that should work—

FEMALE STUDENT: Why should it work if it's wrong, Professor?
(Thinks) Unless, the key is more interesting than the lock?

PROFESSOR: *(Cocking his head to one side)* Well—

FEMALE STUDENT: Shit—that can't be right.

The music rises as the PROFESSOR *runs to hang the chart on the wall, but
he reverses it so the other side is now revealed. Instead of late-model auto-*

mobiles, the other side is illustrated with full-color pictures of prime cuts of meat. All three run to different shelves to pour themselves a drink, but at the last minute, they stop themselves and slap the offending hand reaching for the bottle. They look at the hand for a minute, then shrug and pour themselves a drink anyway. They gasp as it goes down the throat.

PROFESSOR: *(Stepping forward to watch their reactions to the liquor)* No problem. *(He circles back to lean nonchalantly against a bookcase)* Here's another way to look at it. What is a human being—except—that which—

The MALE STUDENT *is staring at the meat chart.*

—dig deep enough now: "Doesn't know."

MALE & FEMALE STUDENT: *(Enlightened, they hit their own foreheads.)* Ahhhhh!

PROFESSOR: Animals, for instance—zebras, lions, buffalos—

Startled, they turn to watch a white screen zip in behind them to stand against the rear wall, a GNOME *behind it. Then the* PROFESSOR *recovers, and slowly approaches the* FEMALE STUDENT.

—and more ordinary things like dogs and cats.

She hasn't been looking at him, but he whirls her around, threateningly. Then he turns from her and picks up the horse, which is still lying on the ground. He carries the horse and stands holding it, as if posing for a photo against the white screen at the rear of the room.

None of those animal things are in a state of not-knowing, because the issue never arises. So a human being is the birth of not-knowing as a real possibility. And when that beneficent stupidity is ended?

He steps forward with the horse. Then he drops it on the floor and returns alone to the white screen.

Then in fact—

The large pink disk rolls in front of him, and he speaks, hidden behind it.

—he, or she, is no longer a human being.

The disk rolls quickly out of the room, revealing the PROFESSOR *facing the white screen with his back to the room. As he takes a few steps backwards, the white screen moves to the side and tilts at a slight angle. The* PROFESSOR *turns back to speak to the room, whispering and slowly kneeling on the collapsed horse.*

Maybe that's why Professor Number Four is taking such a goddamned eternity to park the goddamned automobile.

FEMALE STUDENT: Look, you're into something—*(The white screen whisks off)* I'm just not into, Professor. *(There is a pause. The* PROFESSOR *looks disoriented.)*

MALE STUDENT: Hey, that seems to quiet him down to a considerable extent.

PROFESSOR: I wondered who was going to be the first to say something. *(Looks from one to the other)* What do you know, it was me.

FEMALE STUDENT: You almost said something important, Professor, but it wasn't you.

PROFESSOR: Didn't I speak first?

FEMALE STUDENT: No.

PROFESSOR: Who did?

A Bach-like phrase of music is rising.

MALE STUDENT: I'll have to go into the next room, but I think— *(Points to himself)* "If you go—*(With his other hand, he points to the* PROFESSOR*)*—I go."

PROFESSOR: Really?

MALE STUDENT: Yes.

They all run for a set of cymbal plates with handles, and smack them high up on the wall.

FEMALE STUDENT: *(Calling over her shoulder)* What time is it?

PROFESSOR: My watch stopped! *(He runs out of the room. The two others relax to watch him go.)*

FEMALE STUDENT: *(To the* MALE STUDENT*)* Think about this, Professor. If God himself were to come into this room right now . . . *(She thinks, then turns and places her cymbal plates against his, and with a slight push, sends him spinning from the room. Alone, she faces forward.)* It would mean time had indeed stopped. *(She bangs her two plates together and rubs them against each other.)*

PROFESSOR: *(Offstage, using a Godlike intonation)* God?

FEMALE STUDENT: *(Putting away her plates)* I'm not interested.

PROFESSOR: *(As the* MALE STUDENT *is pushed onstage, wearing a white feather headdress)* You have to be interested. *(He himself appears, wearing a white feather headdress that is much more impressive than his student's.)*

FEMALE STUDENT: My God, that's desperation on your part, Professor.

PROFESSOR: Not on my part.

MALE STUDENT: Not on my part, Professor.

He covers his eyes and tiptoes to a corner, as the others try to hide. It's a game of blindman's buff. The voice on tape is again heard repeating over and over, "Remarkable people never depend on self-revelation."

PROFESSOR: 1, 2, 3, 4, 5, 6, 7, 8, 9, 10. *(Turns from his corner, looks up to the heavens, and starts to spin in place.)* Now, if God, really—

FEMALE STUDENT: Oh, that's vulgar, Professor, and so pretentious, it's doubly vulgar.

MALE STUDENT: *(Shrugging)* Now me—? I'm very naive.

PROFESSOR: Primitive?

MALE STUDENT: Proud of it.

FEMALE STUDENT: What?

PROFESSOR AND MALE STUDENT: Both.

They both collapse to the floor, moaning, "Oh, yeah!" An electronic tone rises, and the FEMALE STUDENT *sits in the chair as a white screen is slid in behind it, giving it the aspect of a throne.*

FEMALE STUDENT: Better sit in this, gentlemen.

PROFESSOR: *(Rising from the floor)* Why?

FEMALE STUDENT: Well, it's a very electric chair. *(She rises from the chair and attaches electric cables to its arms.)*

PROFESSOR: I don't want to be electrocuted, madam.

MALE STUDENT: I don't want to be electrocuted either.

FEMALE STUDENT: Do you see wires?

PROFESSOR: *(Turning away)* There are so many things about these "oh-so-desirable women" I can't justify to myself—

FEMALE STUDENT: You like my company, admit it.

The "busy gnome" music is rising.

MALE STUDENT: True, you like the aggravation, Professor. It energizes.

PROFESSOR: I like that, but I don't like its source, I like its results.

FEMALE STUDENT: It's the same thing.

MALE STUDENT: It's the same thing, Professor.

FEMALE STUDENT: Pleasant dreams, Professor.

She and the MALE STUDENT *run to bookshelves and rip pieces of paper to pieces, throwing them into the air like confetti, as the music rises, along with the voice on the tape repeating, "Remarkable people never depend on self-revelation."*

PROFESSOR: I need lots of sleep!

He takes papers from his briefcase and throws them into the air, as GNOMES *run in to retrieve them, and music and talk all start overlapping in a jumble of sound as the lights grow dim.*

FEMALE STUDENT: No problem—it's as if a curtain were rising.

MALE STUDENT: Imagine a play called "The Pretend Hat." What could happen—

PROFESSOR: *(Overlapping, as he throws his papers)* I'm imagining a play called "Mysteries of Arrogance." That's more my style!

FEMALE STUDENT: Guess!

PROFESSOR: What's a guess?

FEMALE STUDENT: Time will tell!

PROFESSOR: Quite!

FEMALE STUDENT: Counting on internal time, Professor?

PROFESSOR: I turn into somebody who opens his mouth, and whatever comes out travels in desirable directions only!

He has run up against a blackboard, and with the sound of that collision, things quiet down and the GNOMES *run from the room. All that is heard now is an electronic tone. The* PROFESSOR *turns from the blackboard to face the* FEMALE STUDENT, *who is sitting on the floor on the horse, lighting a cigarette. It seems to be late at night. The* PROFESSOR *speaks quietly.*

Automatic truths, madam.

FEMALE STUDENT: Good. What's the technique?

PROFESSOR: *(A pause, then he decides to reveal his secret)* Well, I make up rhymes.

FEMALE STUDENT: Then what?

PROFESSOR: The technique is, they don't rhyme.

FEMALE STUDENT: Really, Professor!

Now there is complete silence.

PROFESSOR: This is supposed to illustrate something.

The FEMALE STUDENT *rises from the floor. The* MALE STUDENT *stands next to the* PROFESSOR, *holding out a man's felt hat.*

FEMALE STUDENT: Ah, your method is the illustrative method.

PROFESSOR: Yes.

FEMALE STUDENT: You make up rhymes that don't rhyme.

PROFESSOR: Not quite. I make up rhymes. I do that. But: they don't rhyme.

FEMALE STUDENT: *(After a pause)* I appreciate the difference.

PROFESSOR: Do you?

The FEMALE STUDENT *takes the felt hat from the* MALE STUDENT *and places it on her closed fist. At the same time, the* PROFESSOR *removes his feather headdress.*

FEMALE STUDENT: Here is my hat.
                        What do you think of that?

PROFESSOR: That rhymes.

FEMALE STUDENT: *(The childlike tune on the piano returns)* Oh. I thought perhaps you'd say to me, That doesn't rhyme.

PROFESSOR: *(He thinks)* But it does.
    "Here is my hat—

He turns to erase the already erased blackboard, and she throws the hat to the floor.

    What do you think of that."

FEMALE STUDENT: Wait a minute. Does that rhyme?

PROFESSOR: We'll have to find out.

FEMALE STUDENT: How?

PROFESSOR: Over the course of time.

He continues to erase, and the MALE STUDENT *steps forward.*

MALE STUDENT: How many people really attend your lecture courses?

PROFESSOR: *(Turning to him in irritation)* Look—! *(Embarrassed to answer, he turns away)* It varies.

MALE STUDENT: Sure.

FEMALE STUDENT: Sure.

MALE STUDENT: But in general, how many?

PROFESSOR: *(Very quietly)* Not many.

FEMALE STUDENT: Come on, how many?

PROFESSOR: Sometimes one. Sometimes two, or three.

He has crossed the room despondently, just as the MALE STUDENT *is reaching for some books on a high shelf, but accidently on purpose makes them tumble to the floor.*

MALE STUDENT: Oh, Jeeze, I'm sorry—another accident.

PROFESSOR: On purpose?

MALE STUDENT: Hey, you know the esteem in which I hold you.

PROFESSOR: *(Sarcastically)* Yeah.

MALE STUDENT: *(As the electronic tone returns)* Therefore the fact that your lectures are not well attended does not, in my eyes, reflect upon you in the least, but rather upon your students.

The FEMALE STUDENT *is quietly rolling the wooden cart into the light. She climbs in, puts her knee up on the railing, studying the* PROFESSOR *from a provocative pose.*

PROFESSOR: You mean, on the ones not totally present.

MALE STUDENT: Possibly those also.

PROFESSOR: But of course—you have no way to know—the quality of those who do attend with regularity.

MALE STUDENT: No.

PROFESSOR: I invite you, Professor. *(He is staring at the* FEMALE STU-DENT.*)*

MALE STUDENT: That's impossible.

PROFESSOR: *(Backs away from the* FEMALE STUDENT*)* My feeling is, the universe uses me as it will. *(Stumbles over the horse and falls to the floor, then looks up in awe)* Jesus, did I just rhyme?

MALE AND FEMALE STUDENT: *(Quietly)* Oh yeah . . . !

PROFESSOR: It's hard to know, but I think it rhymed with something.

MALE STUDENT: I don't think so.

The PROFESSOR *moves to the rear, where he is illuminated by light coming from the next room, and begins talking to himself.*

PROFESSOR: In a certain play entitled "My Head Was a Sledgehammer," a certain character falls deeply in love with his mirror image, although his mirror image doesn't resemble him in many important ways. *(He moves slowly back toward the cart, to gaze at the* FEMALE STUDENT*.)* But is a much more beautiful image. A magic mirror, and the character who has so fallen in love says things that seem beside the point, not expressing love really, but do they?

MALE STUDENT: *(From the shadows at the side of the room)* Do they what?

PROFESSOR: Do they win him the love of women?

MALE STUDENT: Women in general?

The lights have risen, and are bright again.

FEMALE STUDENT: Be more specific.

The PROFESSOR *runs to erase an already erased blackboard and the words pour rapidly out of him, as the two* STUDENTS *reach up and make books tumble from high shelves, and a* GNOME *rolls the cart around the room.*

PROFESSOR: You see what's happening to me? I've been placed in a situation where verbal disorganization—while it does not rule—has been dreamed of deeply by certain individuals who vibrate on the edge of an aura that does divide a particular arena into those who are beautiful and those who are simply pieces of shit.

FEMALE STUDENT: But Professor, isn't that a false distinction?

PROFESSOR: No longer true, madam.

FEMALE STUDENT: Look at ME—?

PROFESSOR: Once again the goddamned philosophers have undone me.

MALE STUDENT: Does this implicate every single one of them?

PROFESSOR: Correct. Every single one of you has cheated me out of my appropriate energies.

MALE STUDENT: Oh, gee. Which of those sons of bitches have done this to you, Professor?

PROFESSOR: I don't know individual names.

MALE STUDENT: Uh-uh! The truth, Professor.

PROFESSOR: Hey—! *(Ominous music begins, and the* PROFESSOR *mutters to himself)* I want it to pour forth abundantly.

MALE STUDENT: What stops you, Professor?

PROFESSOR: Can truth travel between two people? I think not.

FEMALE STUDENT: How come?

The MALE STUDENT *is poised at the blackboard, ready to take notes.*

PROFESSOR: This I can explain. But don't confuse, please, my explanation with the truth.

FEMALE STUDENT: Explain!

The PROFESSOR *runs to his briefcase, takes out a stack of paper, and throws sheets of it into the air. The* GNOMES *pick up pieces of paper and run to the walls, holding the paper flat against the walls as high as their arms can reach.*

PROFESSOR: If the truth is the truth about something—

MALE STUDENT: —But it has to be about SOMETHING!

PROFESSOR: But: that thing it's the truth about—eats it!

MALE AND FEMALE STUDENT: Huhhhh?

The GNOMES *pull away from the walls to watch in disbelief.*

PROFESSOR: And the truth—eaten—disappears by turning into whatever it is that eats it.

MALE AND FEMALE STUDENT: *(Hitting their foreheads in recognition)* Ahhhh!

The GNOMES *run out of the room.*

PROFESSOR: In order to clarify this, I shall now tear to pieces an important envelope, within which—*(The cabinet rolls in quickly. The* PROFESSOR *runs to it, takes out an envelope and opens it and reads a piece of paper, which he then tears to pieces and throws into the air like confetti.)* I forgot what was in the envelope. But that's okay—because it illustrates my real and most secret import.

FEMALE STUDENT: *(Rubbing her body up against him)* I don't know how to say this, Professor. But what you just did really turns me on.

PROFESSOR: Not really?

FEMALE STUDENT: Really.

The MALE STUDENT *has also come to cuddle up against the* PROFESSOR.

PROFESSOR: You're not lying to me?

FEMALE STUDENT: Truth is—I'm not lying.

Irritated, the PROFESSOR *slaps the* MALE STUDENT *to make him back off. Then he runs to a bookcase and tears up more paper, which he throws into the air.*

PROFESSOR: How could I possibly know who's lying?

A doorbell rings.

FEMALE STUDENT: My God, that must be me.

PROFESSOR: *(Calls out, as lively dance music is beginning)* Just a minute!

Both he and the MALE STUDENT *find more paper to tear up and throw into the air. As it floats down, they shake hands, then bump shoulders and go into the dance they have already performed where they use their hands in a pumping motion, swing their hips in place, and then do aggressive breast strokes as they strut around the stage. The* FEMALE STUDENT *picks up the horse, holds it between her legs, and tries to ride the horse across the stage in time to the music. Then she falls with a scream.*

FEMALE STUDENT: Why did you make me do this, Professor?

PROFESSOR: *(As the dance stops)* I don't know of course.

FEMALE STUDENT: Intuition?

PROFESSOR: I can't say it was intuition.

The dance recommences minus the FEMALE STUDENT *as two* GNOMES *enter carrying a giant white envelope high in the air, marching in time to the music. The* MALE STUDENT *breaks from the dance to grab a small envelope attached to the big one and stares at it, as the dance music gives way to a return of the childlike piano tune and the big envelope is carried off.*

MALE STUDENT: Hey, what can you say about what you've been doing since you came into this room, Professor?

PROFESSOR: Theorize about it? Spare me—

MALE STUDENT: I'm afraid there's no other option, Professor. *(Holds the small envelope to his own forehead)* Close your eyes!

FEMALE STUDENT: What a shame, Professor. *(She is crossing the room carrying an egg she holds over a mixing bowl)* You haven't produced the desirable egg. And your heart, Professor, which was never whole, breaks—

She breaks the egg into the bowl, as the MALE STUDENT, *who has torn his envelope to pieces, throws it into the air.*

MALE STUDENT: Hey, what effect is this having on you, Professor?

PROFESSOR: Remarkable enough, it brings tears to my eyes.

MALE STUDENT: Genuine magic?

FEMALE STUDENT: No tears visible. He makes you believe in them simply by his tone of voice. *(Runs to his side.)*

PROFESSOR: Are you threatening to saw me in half, madam?

MALE STUDENT: She doesn't know how to do that without hurting you, Professor.

PROFESSOR: Try hurting me. Go ahead, twist something.

Each twists one of his arms, and he grunts in pain.

Now I'm feeling a little pain. Though it would be more poignant doing this in public.

They twist him around and slam him into a blackboard, as the VOICE *on tape is heard repeating again and again, "Remarkable people never depend on self-revelation" softly through the following conversation.*

What am I able to perceive, even through tears of pain?

MALE STUDENT: Bypass the tears, move to the ears.

PROFESSOR: What's happening?

MALE STUDENT: Oh, it hasn't happened Professor, so that's why it's still interesting.

PROFESSOR: What is?

MALE STUDENT: Well, you've been crying.

A music of anticipation is heard.

PROFESSOR: Nobody noticed—

FEMALE STUDENT: Don't believe him. He was crying real tears.

MALE STUDENT: That's what I thought—

PROFESSOR: Don't believe it—

The cabinet rolls quickly into the room and he runs up to it.

My next trick? This book disappears. *(Takes out a book and holds it up in the air.)*

MALE STUDENT: Was the drawer empty?

PROFESSOR: Careful. I make things appear out of nothing.

He throws the book back into the cabinet, and he and the MALE STUDENT *circle each other with their hands up in the air as if to say, "Nothing up my sleeve."*

MALE STUDENT: Bravo, Professor!

PROFESSOR: Thank you, Professor!

They run toward each other, and just before they collide, fall suddenly on their knees and roll away from each other. Then they fall back exhausted onto the floor.

MALE STUDENT: Now, tell me, Professor, why do you prefer the disappearing act to the opposite?

PROFESSOR: That takes some explaining—

FEMALE STUDENT: Both kneel for this.

PROFESSOR: *(Rising)* Unnecessary.

FEMALE STUDENT: I assume you both kneel for this, more than once, please.

PROFESSOR: *(Muttering to himself)* Perhaps I have been waiting to enter this very arena.

The FEMALE STUDENT takes a stack of books from the bookcase, turns back to the PROFESSOR, and deliberately throws the books onto the floor. At the moment of impact, the PROFESSOR and the MALE STUDENT hold their heads and collapse slowly, moaning, "Oh, yeah!"

FEMALE STUDENT: Should I repeat myself?

MALE STUDENT: Does she mean repeating myself even when I don't remember?

FEMALE STUDENT: Everybody—out of this room, please!

MALE STUDENT: What did I say?

FEMALE STUDENT: *(Momentarily perplexed)* If I say everybody—am I included?

MALE STUDENT: Do it! Do it immediately, please.

The FEMALE STUDENT runs out of the room and the MALE STUDENT comes forward on his hands and knees, unaware of the PROFESSOR behind him.

Now that I can't see anybody because they left the room, I can't see anybody, because they left the room.

PROFESSOR: Amazing, sir.

MALE STUDENT: Huh?

PROFESSOR: Your reality is my own reality.

MALE STUDENT: If that's true—?

PROFESSOR: A particular friend helped me out of an impasse.

MALE STUDENT: How?

PROFESSOR: He, or she, blocked all my escape routes.

The MALE STUDENT *turns to get away but bumps into a wall and spins off against a blackboard, as an electronic tone rises.*

MALE STUDENT: But was it a he or a she?

PROFESSOR: I can't possibly remember.

MALE STUDENT: The truth, Professor—

PROFESSOR: That's what I meant from the beginning. I can't remember!

The tone rises momentarily, as the pink disk rolls in, and entering with it, and posing in front of it, is the FEMALE STUDENT, *who is now wearing an exotic bright red shoe on one foot.*

FEMALE STUDENT: What changed the minute I left the room, Professor? Come on, tell me what's radically different.

MALE STUDENT: It always happens like this. As soon as someone asks me, What's radically different, if I fail to notice what's radically different, I find myself in very deep shit.

PROFESSOR: I find myself in very deep shit, Professor.

FEMALE STUDENT: Finally, somebody noticed.

PROFESSOR: What was noticed?

FEMALE STUDENT: I bought new shoes, Professor. *(She grabs a leg of the horse, swings it around, and positions herself with one foot displayed on the horse's rump)* Or—at least one.

MALE STUDENT: Jesus Christ, the one thing I failed to notice was an individual shoe!

PROFESSOR: I think we're both in deep shit, Professor.

FEMALE STUDENT: Dizzy? Me too.

She runs excitedly forward and spins herself, then screams and falls, as the GNOMES *enter and pull strings from the wall and dangle the little balls over the floor near her body.*

PROFESSOR: This is too much. The angel who dominated my life turns human, but do I believe it?

FEMALE STUDENT: I bought new shoes, Professor, that's all.

PROFESSOR: Why? Of all possibilities—?

FEMALE STUDENT: Simple. To get back down to earth.

Ominous music plays softly.

PROFESSOR: Run that by me again.

FEMALE STUDENT: *(Rising from the floor)* Careful, Professor. When I pick up speed, I pick up more than speed.

PROFESSOR: Does it really look like I'm about to gravitate toward whatever pulls me in directions I'm totally incapable of traveling?

FEMALE STUDENT: *(Trying to figure things out)* The only thing I can come up with: Somebody must have had God in mind—remember him?—when he went down on his hands and knees in front of me.

MALE STUDENT: That was last night, madam. *(Going down on his hands and knees before her)* But here we have a whole new beginning.

FEMALE STUDENT: I don't see the difference.

PROFESSOR: *(Also going down on all fours)* Here's the difference.

FEMALE STUDENT: *(Circling the two men, as the* GNOMES *sneak out of the room)* Right again, Professor. The effect is totally different.

PROFESSOR: Continue, please.

FEMALE STUDENT: Well, the word that comes to mind immediately is . . . *(Thinks for a while)* Literature?

PROFESSOR: Stop groping for respectability, madam.

FEMALE STUDENT: I was doing the opposite, Professor. If I say "literature," that's like saying "shit."

The GNOMES *enter, tearing pages from books and scattering them over the floor.*

Whereas if I started rolling my tongue over an objective word like "science"? Well, that's sweet enough to plunge somebody like me into those nether regions where pigs do fly, Professor.

PROFESSOR: Strange, to me science has no particular aroma.

The FEMALE STUDENT *kicks his behind, and he jumps up from the floor.*

Ow! Why did you do that?

FEMALE STUDENT: Figure it out, my friend.

The PROFESSOR *comes down to her and looks into her eyes.*

PROFESSOR: I'm your friend no longer, madam.

He goes, but she grabs his arm to stop him, as a GNOME *is rolling out a thin red carpet diagonally across the floor, and sad violin music is heard.*

FEMALE STUDENT: No longer the friend of science?

The PROFESSOR *pulls away and leaves the room.*

I must have fallen prey to the unenviable error of being reticent when it comes to putting my best foot forward.

The MALE STUDENT *has picked up the horse and carries it down to her.*

MALE STUDENT: Maybe it's time to try genuine role reversal?

FEMALE STUDENT: Just as an experiment?

MALE STUDENT: Look at it this way. If it's an experiment, we can pretend nobody really gets hurt. *(He throws her the horse, which she catches in her arms, as the pink disk rolls out of the room and is simultaneously replaced by the cabinet, which now offers a display of shoes. The* MALE STUDENT *selects a pair with red apples attached to the tips of the toes. He sits on the floor putting them on.)*

FEMALE STUDENT: Inconclusive. This is inconclusive. *(She throws away the horse, as the childlike piano tune returns.)*

MALE STUDENT: *(Standing up and looking sadly at his shoes)* As a matter of fact, this is inconclusive.

FEMALE STUDENT: *(Joining him in contemplation)* But it still counts. Because our agreement is—it's inconclusive.

She runs back to grab the horse, pulls it along the red carpet to one end, and sits down on top of it, as the MALE STUDENT *lifts the apples off his shoes and studies them.*

MALE STUDENT: That reminds me—

FEMALE STUDENT: What?

Two GNOMES have appeared and are kneeling, each holding out a loaf of french bread toward the MALE STUDENT. Each bread has a little napkin draped over its middle.

MALE STUDENT: *(Backing away from the bread)* Wait a minute. Am I being asked to invent something?

He reaches for a book on a high shelf, perhaps to research some important invention, but he accidently tumbles books down over himself. At the same moment the GNOMES are permanently fixing an erect french bread on the toe of each of his shoes. The tiny napkins now delicately cover the upright tips of the bread. The MALE STUDENT takes a few tentative steps, studying the bread that rises from his feet.

MALE STUDENT: You're not going to believe this.

FEMALE STUDENT: Try me, Professor.

MALE STUDENT: This might not be an appropriate moment, but— *(Bends toward the bread)*—I'm really hungry!

FEMALE STUDENT: Ah, just like Professor Number One, Professor Number Two, Number Three, Number Four, Number Five, Number Six—

MALE STUDENT: Wait a minute.

FEMALE STUDENT: What are you trying to tell me?

MALE STUDENT: How do I know? *(He is lifting a bread from one shoe, holding it close to his body so it points out from between his legs toward the FEMALE STUDENT, who averts her face.)*

FEMALE STUDENT: Then how can I share your experience?

MALE STUDENT: Oh, I don't see how it can be avoided.

FEMALE STUDENT: It's being avoided.

MALE STUDENT: Let me show you something.

FEMALE STUDENT: What?

MALE STUDENT: *(Extends the bread a bit more)* You decide.

FEMALE STUDENT: I choose not to see this.

MALE STUDENT: Ahh—

FEMALE STUDENT: *(Whirls to face him, pointing at the bread)* I choose not to see this thing.

MALE STUDENT: —I don't imagine this bread appeals to a female professional such as yourself.

FEMALE STUDENT: Of course not. It's much too long. *(Turns and throws her horse away.)*

MALE STUDENT: But. Here's the thing—shorten it.

FEMALE STUDENT: How?

MALE STUDENT: *(His eyes glistening)* Hey, use a knife.

FEMALE STUDENT: You do that, please. *(Hides her eyes and runs to a far corner.)*

MALE STUDENT: With a genuine—knife?

FEMALE STUDENT: With a knife!

MALE STUDENT: *(Backing up along the carpet, caressing his bread)* Ah, with a genuine knife?

FEMALE STUDENT: *(Grabs the bread from him and throws it away. Ceremonial horns are heard honking.)* I said with a knife, because what I intended, was a knife!

They both run to the bookcases and throw down even more books, searching for knives, which they find hidden behind the books. They hold them up into the light, which is now intense, and white streamers attached to the knife handles flutter as they move the knives in circles. The music builds and the white hospital screen is rolled into the room by two GNOMES. *Behind the screen is the* PROFESSOR *in his feather headdress, but he towers above it because he is walking on tall cothurni. The screen pivots around him as he slowly advances, and he is seen supporting himself with a long staff. In his other hand is a knife held high in the air, with streamers falling from its handle. Strapped to his waist is an open book trailing long streamers. The music changes to the childlike piano tune. The* PROFESSOR *advances slowly, his feet weighed down by the cothurni.*

MALE STUDENT: Hey, Professor! Why do you walk that way?

PROFESSOR: *(Staring up into the light)* When you get old, Professor—that's what it feels like. So, my God, get used to it, Professor!

Two GNOMES *cause the screen to circle him slowly.*

Because that's what it's going to feel like, when your head pivots in the direction of all desirable ladies, and you WOBBLE, Professor—you too Professor! And if you don't get ready for that, Professor, you're going to be very, very unhappy! *(The screen is in front of him now, and he cries out, lifting his arms in the air)* Ohhh! Very unhappy!

The ominous music becomes deafening, while the GNOMES *are tearing pieces of paper to bits, which they throw into the air as the* MALE STUDENT *and the* FEMALE STUDENT *run after the floating pieces of paper, trying to stab them with their knives. Then the loud music abruptly stops, and everything is silent, and then a wistful violin is heard, and the* PROFESSOR *lifts his face up over the edge of the hospital screen.*

PROFESSOR: *(Very quietly)*
See what happens?
Time passes through unchartable waters.
Language proliferates
just so things can be said nobody ever intended.
Then,
they come true.

FEMALE STUDENT: I want to see this.

PROFESSOR: I speak. It comes true.

MALE STUDENT: *(Advancing toward the screen)* How, Professor?

PROFESSOR: Pure speed, Professor.

MALE STUDENT: Speed?

PROFESSOR: Jumping fast over every new consideration. *(Points a finger toward the* FEMALE STUDENT*)*
Bang!
And beyond that—nothing.
Getting there really fast, Professor:
Then, nothing.

MALE STUDENT: Hey, that sounds like skipping even, well, life itself, Professor.

He is circling the room as the GNOMES *slowly remove the screen from around the* PROFESSOR *and install it at the rear of the room, in the shadows, where the* GNOMES *hide behind it.*

PROFESSOR: Right, everything so fast, life is skipped. *(Proceeding slowly down the red carpet)* Right.

From behind the screen, the GNOMES *throw their hands up in the air. Just their wiggling hands are visible as they shout gleefully, "Wheee!"*

MALE STUDENT: Now, wait a minute, what's achieved if life is simply skipped, Professor?

PROFESSOR: Nothing has to be achieved. That's the beauty of it.

MALE STUDENT: Everything skipped? Then one arrives at nothing, Professor.

He gestures toward the FEMALE STUDENT, *who cracks open an egg over a mixing bowl. Again the* GNOMES *throw up their hands and shout "Wheee!"*

Why is that desirable?

PROFESSOR: I didn't say it was desirable.

MALE STUDENT: Hey! What is it, if it isn't desirable, Professor?

PROFESSOR: Necessary, that's all.

MALE STUDENT: Necessary for what?

PROFESSOR: I don't know. You, you don't know. *(Lifts his staff and points it toward the* FEMALE STUDENT*)* She, she, she doesn't know—

Music rises as the MALE STUDENT *and* FEMALE STUDENT *join hands and whirl once as the* GNOMES *shout, "Whee!"*

MALE STUDENT: Then slow down, Professor. Enjoy the ride at least!

As the STUDENTS *come out of their spin, all but the* PROFESSOR *are whirled out of the room. He is left alone in a shaft of light, and he laboriously crosses the room on his cothurni. It is very quiet, just the faraway sound of the childlike piano tune.*

PROFESSOR: *(To himself)* Why do you suppose it is that whenever I remember you, you seem very uncomfortable, Professor? While I, after all is said and done, seem perfectly comfortable.

MALE STUDENT: *(His voice only, heard over the loudspeakers)* Suppose I left, forever. You'd be bored.

Still alone, the PROFESSOR *looks with anguish to the sky, and spreads his arms, as if pleading for mercy. The* MALE STUDENT *enters slowly, but the* PROFESSOR *doesn't seem to be aware of him.*

You'd be all alone.

PROFESSOR: *(Turning slowly to look at him)* It wouldn't matter.

FEMALE STUDENT: *(Entering the room)* If you were all alone, that would matter, Professor.

PROFESSOR: Oh no it wouldn't. It wouldn't even be part of my life. Because life—that's not where it's happening, madam. You think it's happening in life: you're wrong, beautiful madam.

FEMALE STUDENT: Well, for instance . . . what other place could it be happening?

PROFESSOR: *(Moves laboriously away from her)* No, it's not happening someplace else. It's happening right here. But it's not happening in life.

FEMALE STUDENT: Where is "here," Professor, that isn't in life, please? And if it's not it's someplace else?

PROFESSOR: *(Turning on her in irritation)* Well, why don't we just shut up and admit I'm right?

The pink disk rolls into the room, hiding the MALE STUDENT.

FEMALE STUDENT: But that means—give up total power over my own mind.

PROFESSOR: Okay. That means powerlessness. Bye-bye, Professor—

He points his staff toward the disk, which rolls out of the room, taking the MALE STUDENT *with it.*

Ah, maybe that's the point.

FEMALE STUDENT: Somebody's mind is a total blank.

PROFESSOR: Ah, isn't that the point, madam?

FEMALE STUDENT: What?

PROFESSOR: *(Listening to a strange new music, looking around the entire room)* What is this?

FEMALE STUDENT: Among myriad possibilities, I don't know to what you refer.

PROFESSOR: *(As the male student sneaks back into the room)* This is an idea that is not allowed to come into existence. *(Both the* MALE STUDENT *and* FEMALE STUDENT *slowly sink to the floor, muttering to themselves "Oh yeah . . . !")* Its power as an idea is not thereby minimized. *(He has turned his back and is slowly walking to the rear wall)* Ohhhhh—it still exerts power.

MALE STUDENT: *(From the floor)* Hey . . . what idea is that?

PROFESSOR: *(Turning back to them)* This is an idea that doesn't exist. It makes the walls of this room vibrate. And failing to do so, something else writes on my forehead, letter of invisible fire.

MALE STUDENT: What I don't see is the fire.

PROFESSOR: It's invisible fire!

MALE STUDENT: Sorry, Professor—if it's invisible, it's not in life. And if it's not in life, I'm not interested, because I'm here in life.

He and the FEMALE STUDENT *lift their arms and start slowly spinning through the room, as a new music rises.*

So what interests me is here, where I am.

He and the FEMALE STUDENT *meet and tango together for a few steps, then break away from each other and spin to opposite walls as the* PROFESSOR *shouts over the music.*

PROFESSOR: Hey, that's silly, Professor!

FEMALE STUDENT: Why, Professor?

PROFESSOR: Because what's interesting, Professor, is what's there— *(Pointing his staff at the* MALE STUDENT*)*—where you are!

MALE STUDENT: Here I am!

PROFESSOR: But—! That you don't know!

MALE STUDENT: Don't know what?

A pause, and the PROFESSOR *turns away.*

PROFESSOR: Where you are.

MALE STUDENT: Ah, but even if I don't know about it, that's still here in life, Professor.

FEMALE STUDENT: Hello again.

PROFESSOR: Once again, very silly.

MALE STUDENT: Do better than that, Professor.

PROFESSOR: *(As a pounding music begins to rise, he stomps across the room.)* Okay. That's unalterably SILLY, Professor!

MALE AND FEMALE STUDENT: *(Collapsing to the floor)* Ohhh yeah . . .

The pounding music rises, as the GNOMES *roll in four large disks with blackboards attached, and begin to frantically erase the blackboards.*

PROFESSOR: —because the most interesting thing about anybody in this room is where you are! That part of where you are that isn't where you think you are!

FEMALE STUDENT: Oh, my God—does that rhyme, Professor?

Both STUDENTS *have run to blackboards and are frantically erasing.*

MALE STUDENT: That's silly.

FEMALE STUDENT: That rhymes!

MALE STUDENT: That's so silly!

PROFESSOR: Now that's REALLY silly—

All the others, GNOMES *included, join hands, shouting, "Wheee!" as they circle once around the* PROFESSOR *and then return to their work at the blackboards.*

Then again, if you really think that's silly—?

He is bellowing through the pounding music when a flash of light makes everyone erasing fall to the ground and then immediately jump up to start erasing again. Then another flash of light, and again they fall and jump back to work—and through all this chaos the PROFESSOR *is striding across the stage on his cothurni and shouting over the music.*

Better think it through again!
From the beginning!
One more time! *(He beats the air with his staff, in rhythm with his invocation)*
One more time!
One more time!

The others have fallen to the floor one last time, and stay frozen with their legs up in the air, as the PROFESSOR *swings his staff into the air one last time and doesn't move, as the light suddenly goes out and the music stops.*

THE END

Chronology

1937	Born June 10, New York City.
1959	B.A. degree, Magna Cum Laude, Brown University.
1962	M.F.A. degree, Playwriting, Yale Drama School.
1968	Founds Ontolological-Hysteric Theater.
	Angelface, Ontological-Hysteric Theater at Jonas Mekas's Cinemateque, NYC.
1970	*Elephant Steps* (Silverman), libretto, Berkshire Music Festival, Tanglewood, Mass.; Hunter Opera Theatre, NYC.
	First of ten *Village Voice* Obie awards.
1971	*Dream Tantras for Western Massachusetts,* libretto, Music-Theatre Group, Lenox Arts Center, Lenox, Mass.
	Total Recall, Ontological-Hysteric Theater, Cinemateque, NYC.
1972	*Evidence,* Ontological-Hysteric Theater, Theater for a New City, Bank Street, NYC.
	HcOhTiEnLa or Hotel China, Ontological-Hysteric Theater, Cinemateque, NYC.
	Dr. Selavy's Magic Theatre (Silverman), libretto, Lenox Art Center, Lenox, Mass.; Mercer-O'Casey Theatre, NYC, 1972–73.
1973	*Particle Theory,* Ontological-Hysteric Theater, Theater for a New City, NYC.
	Classical Therapy, or A Week under the Influence, Festival d'Automne, Théâtre Recamier, Paris.
	Sophia = (Wisdom): Part 3: The Cliffs, Ontological-Hysteric Theater, Theater for a New City, NYC.
1974	*Pain(T)* and *Vertical Mobility,* presented in repertory, Ontological-Hysteric Theater, 141 Wooster Street, NYC.

237

Hotel for Criminals (Silverman), libretto, Lenox Art Center, Mass.; Music Exchange Group, NYC, 1975.

1975   Ontological-Hysteric Theater moves to its new theater at 491 Broadway.

Pandering to the Masses: A Misrepresentation, Ontological-Hysteric Theater, 491 Broadway.

Out of Body Travel, video play.

1975–76   *Rhoda in Potatoland,* Ontological-Hysteric Theater, 491 Broadway.

1976   *Threepenny Opera* (Brecht-Weill), Vivian Beaumont Theatre at Lincoln Center; Delacorte Theatre, Central Park.

Livre des Splendeurs, Festival d'Automne, Les Bouffes du Nord, Paris.

1977   *Book of Splendors, Part Two,* Ontological-Hysteric Theater, 491 Broadway.

Blvd de Paris: I've Got the Shakes, Ontological-Hysteric Theater, 491 Broadway.

City Archives, video.

1978   *Stages* (S. Ostrow), Belasco Theatre.

1978–79   *Strong Medicine,* feature film, produced by Ontological-Hysteric Theater.

1979   Foreman sells theater at 491 Broadway.

1980   *Madame Adare* (Silverman), libretto, New York City Opera.

Luogo + Bersaglio (Place + Target), Rome, Milan, and Turin.

1981   *Penguin Touquet,* Public Theater, NYC.

Don Juan (Molière), Guthrie Theatre, Minneapolis; Delacorte Theatre, Central Park, 1982.

Café Amerique, Festival d'Automne, Paris and European tour, with performances at Strasbourg, Grenoble, Villeurbanne, Turin, Nice, and Amsterdam.

1982   *Three Acts of Recognition* (B. Strauss), Public Theater, NYC.

Dr. Faustus Lights the Lights (Stein), Festival d'Automne, Paris; Berlin Festival.

1983   *Die Fledermaus* (R. Strauss), Paris Opera, France.

Egyptology, Public Theater, NYC.

La Robe de Chambre de Georges Bataille, Festival d'Automne, Théâtre de Gennevilliers, Paris.

The American Imagination (Silverman), libretto, Music-Theatre Group, NYC.

1984   *The Golem* (Levick), Delacorte Theatre, Central Park.

        The Birth of a Poet (Acker, Gordon, Salle), RO Theatre, Rotterdam; Next Wave Festival, Brooklyn Academy of Music, 1985.

        Dr. Selavy's Magic Theatre (Silverman), revival, Music-Theatre Group, St. Clement's Church, NYC.

1985   *Ma Mort, Ma Vie, de Pier Pasolini* (Acker), Théâtre de la Bastille, Paris.

        Miss Universal Happiness, Ontological-Hysteric Theater and Wooster Group, Performing Garage, NYC.

1986   *Africanus Instructus* (Silverman), Music-Theatre Group, St. Clement's, NYC.

        Largo Desolato (Hável), Public Theater, NYC.

        The Cure, Performing Garage, NYC.

        End of the World with Symposium to Follow (Kopit), American Repertory Theatre, Cambridge, Mass.

1987   *Film Is Evil: Radio Is Good,* Ontological-Hysteric Theater and NYU Tisch School of the Arts, NYC (including *Radio Rick in Heaven and Radio Richard in Hell,* film).

        Love & Science, Theatre Aurora, Stockholm, Sweden; Stockbridge, Mass., 1990.

1988   Sustained Achievement Award for Twenty Years in the Theatre, *Village Voice* special Obie.

        Symphony of Rats, Ontological-Hysteric Theater and Wooster Group, Performing Garage, NYC.

        The Fall of the House of Usher (Glass / Yorinks), American Repertory Theatre, Cambridge, Mass., and Louisville Opera, Ky.; reconceived version, Teatro Communale, Florence, Italy, 1992.

        What Did He See? Public Theater, NYC.

1989   *Where's Dick?* (Wallace / Korie), Houston Grand Opera.

        Lava, Ontological-Hysteric Theater and Wooster Group, Performing Garage, NYC.

1990   NEA Distinguished Artist Fellowship for Lifetime Achievement in Theatre.

        Woyzeck (Büchner), Hartford Stage Company, Conn.

        Eddie Goes to Poetry City (Part 1), Ontological-Hysteric Theater and New City / Theatre Zero, Seattle, Wash.

        Total Rain, video.

1991   *Eddie Goes to Poetry City (Part 2),* La MaMa Annex, NYC.

1992   Ontological-Hysteric Theater moves to St. Mark's Church, NYC.

        The Mind King, Ontological-Hysteric Theater at St. Mark's.

1993   Honorary Doctorate, Brown University.

        Samuel's Major Problems, Ontological-Hysteric Theater at St. Mark's.

1994   *My Head Was a Sledgehammer,* Ontological-Hysteric Theater at St. Mark's.

1995   *I've Got the Shakes!* Ontological-Hysteric Theater at St. Mark's.

        Don Giovanni (Mozart / Da Ponte), Opéra de Lille, France.

1996   MacArthur Fellowship.

        Edwin Booth Award for Theatrical Achievement, City University of New York.

        The Universe, Ontological-Hysteric Theater at St. Mark's.

        Venus (Parks), Public Theater, NYC.

1996–97 *Permanent Brain Damage,* Ontological-Hysteric Theater at St. Mark's and Meltdown Festival, London.

        Pearls for Pigs, Hartford Stage and on a national and international tour; Tribeca Performing Arts Center, NYC, 1997.

1997   *Benita Canova,* Ontological-Hysteric Theater at St. Mark's.

1999   *Paradise Hotel,* Ontological-Hysteric Theater at St. Mark's.

Select Bibliography

Note: Reviews are too numerous for inclusion here. Those further interested in Foreman's critical reception should match the chronology with indexes of contemporary newspapers and journals.

Books by Richard Foreman

Plays and Manifestos. Edited by Kate Davy. New York: New York Univ. Press, 1976.

Warum ich so gute Stücke schreibe. Berlin: Merve Verlag, 1982.

Reverberation Machines: The Later Plays and Essays. Barrytown, N.Y.: Station Hill Press, 1985.

Love & Science: Selected Music-Theatre Texts. New York: Theatre Communications Group, 1991.

Unbalancing Acts: Foundations for a Theatre. New York: Pantheon Books, 1992.

My Head Was a Sledgehammer: Six Plays. Woodstock, N.Y.: Overlook Press, 1995.

No-Body: A Novel in Parts. Woodstock, N.Y.: Overlook Press, 1997.

Articles by Richard Foreman

"Critique: Glass and Snow." *Arts Magazine,* February 1970, 20–22.

"The Future of the Theatre: The Theatre Falls, Accurately to Pieces." *Confrontation,* no. 11, 1975, 145–48.

"Quatorze Observations." *Tel Quel* (Paris), no. 68 (January 1976), 57–64. Original English manuscript printed as "14 Things I Tell Myself," in *Reverberation Machines,* 222–30.

"How to Write a Play." *Performing Arts Journal* 1 (fall 1976): 84–92. Reprinted in *Reverberation Machines,* 222–30.

"The Carrot and the Stick." *October* 1, no. 1 (1976): 22–31. Reprinted in *Reverberation Machines,* 214–21.

"How I Write My (Self: Plays)." *TDR* 21 (December 1977): 5–24. Reprinted in *Reverberation Machines,* 231–42.

"The Life and Times of Sigmund Freud." *Village Voice,* September 4, 1978, 75.

"How Truth . . . Leaps (Stumbles) across the Stage." *Performing Arts Journal* 14 (1981): 91–97. Reprinted in *Reverberation Machines,* 198–203.

"Die Mütter von Uns allen: Richard Foreman über Gertrude Stein." *Theater Heute* 23, no. 10, October 1982, 36–37.

"The Art of Kathy Acker." *On the Next Wave* 3, no. 3 (November 1985): 24.

"During the Second Half of the Sixties." In *To Free the Cinema.* Edited by David James. Princeton, N.J.: Princeton Univ. Press, 1992, 138–44.

"Ages of the Avant-Garde." *Performing Arts Journal* 46 (January 1994): 15–18.

"Awful Great: On Jack Smith." *Artforum,* October 1997, 74.

Interviews and Symposia with Richard Foreman

"Entretien avec Richard Foreman." By Guy Dumur. *Chroniques de l'art vivant* (Paris), October 1, 1973, 31–32.

"An Interview with Richard Foreman." By Michael Feingold. *Yale / Theater* 7, no. 1 (1975): 5–29.

"Les faux pas de la logique." By Colette Godard. *Le Monde* (Paris), September 16, 1976, 15.

"From Soho to Paris and Back Again." By Theodore Shank. *Soho Weekly News,* January 27, 1977, 17–19.

"Art in the Culture." Symposium with John Cage, Richard Foreman, and Richard Kostelanetz. *Performing Arts Journal* 11 (summer 1979): 71–84.

"Through Cinema to Cinema: Conversation with Richard Foreman." By John Hagen. *Millennium Film Journal,* no. 3, winter 1979, 5–24.

"Le nostre parole al centro del bersaglio." By Nico Garrone. *La Repubblica* (Rome), December 1, 1979, 15.

"Hören + Sehen. Wohin das alles zielt." By Helga Finter. *Theater Heute* (Hannover) 21, no. 9, September 1980, 26–27.

"Writing and Performance." Symposium with Richard Foreman, Richard Kostelanetz, Linda Mussman, and Robert Wilson. *New York Arts Journal* 28 (December 1982): 25–26.

"Richard Foreman: Down and Up in Paris." By Timothy na Gopaleen. *Village Voice,* December 27, 1983, 107.

"The Theatre of Richard Foreman." By Roger Oliver. *On the Next Wave* 3, no. 3 (November 1985): 16–23.

"Both Halves of Richard Foreman: The Playwright." By David Savran. *American Theatre,* August 1987, 14, 19–21, 49–50. Reprinted in *In Their Own Words.* New York: Theatre Communications Group, 1988.

"Both Halves of Richard Foreman: The Director." By Arthur Bartow. *American Theatre,* August 1987, 15–18, 49. Reprinted in *The Director's Voice.* New York: Theatre Communications Group, 1988.

"Bouncing Back the Impulse." By Nick Kaye. *Performance* (London), September 1990, 31–42.

"Off-Broadway's Most Inventive Directors Talk about Their Art." By Elizabeth LeCompte. *Village Voice,* August 10 / 16, 1994, 29–34.

"Richard Foreman." By Eric Bogosian. *Bomb,* spring 1994, 30–34.

"Today I Am a Fountain Pen." By Elinor Fuchs. *Theater* (Yale) 25, no. 1 (spring / summer 1994): 82–86.

"Beyond Sense and Nonsense: Perspectives on the Ontological at 30." Symposium with Charles Bernstein, Arthur Danto, Richard Foreman, Sylvère Lotringer, and Annette Michelson. *Theater* (Yale) 28, no. 1 (1997): 23–34.

"Like First-Class Advertising." By Arnold Aronson. *Brecht Yearbook* 23, October 1997, International Brecht Society / Berliner Ensemble, 88–92.

"More Hysteria Please: A Psychoanalytic Session." By Josefina Ayerza. *Lacanian Ink,* no. 12, fall / winter 1998, 14–37.

Interviews with Kate Manheim

"Theme and Variation within an Erotic Landscape." By Florence Falk. *SoHo Weekly News,* December 22, 1977, 23–25.

"Lavorare con Foreman." By Ruggero Bianchi. *Scena* (Milan), no. 5–6, December 1979, 32–34.

"Kate Manheim: 'I Don't Need No Teacher.'" By Ellen Rapp. *Village Voice,* August 16, 1983, 80–81.

"Kate Manheim ne comprends rien à Richard Foreman." By Anne Laurent and Jean Kalman. *Libération* (Paris), September 30, 1983, 28.

"Talking with Kate Manheim: Unpeeling a Few Layers." By Richard Schechner. *TDR* 31 (winter 1987): 136–42.

Books on Richard Foreman

Benmussa, Simone, ed. *Le théâtre de Richard Foreman.* Paris: Gallimard, 1973.

Bigsby, C. W. E. *A Critical Introduction to Twentieth-Century American Drama.* Vol. 2, *Beyond Broadway.* New York: Cambridge Univ. Press, 1985.

Cohn, Ruby. "Visions and Visuals." In *New American Dramatists, 1960–90,* 2d ed. New York: St. Martin's, 1991, 145–53.

Cole, Susan. *Directors in Rehearsal.* New York: Routledge, 1992.

Davy, Kate. *Richard Foreman and the Ontological-Hysteric Theatre.* Ann Arbor: UMI Research Press, 1981.

Kaye, Nick. *Postmodernism and Performance.* London: Macmillan, 1994.

Léonardini, Jean-Pierre, ed. *Festival d'Automne à Paris 1972–1982.* Paris: Messidor, 1982.

MacDonald, Erik. *Theatre at the Margins.* Ann Arbor: Univ. of Michigan Press, 1993.

Marranca, Bonnie. *The Theatre of Images.* New York: Drama Book Specialists, 1977.

Pasquier, Marie-Claude. *Le théâtre américain d'aujourd'hui.* Paris: Presses Universitaires de France, 1978.

Quadri, Franco. *Il Teatro degli Anni Settanta: Invenzione de un Teatro Diverso.* Turin: Giulio Einaidi, 1984.

Robinson, Marc. *The Other American Drama.* New York: Cambridge Univ. Press, 1994.

Shank, Theodore. *American Alternative Theatre.* New York: Grove Press, 1982.

Articles on Richard Foreman

Aronson, Arnold. "Richard Foreman as Scenographer." *TheatreForum,* winter / spring, 1997, 17–23.

Bode, Walter. "Richard Foreman." *Contemporary Dramatists.* Edited by K. A. Berney. London: St. James Press, 1994, 163–67.

Brustein, Robert. "Theatre with a Public Dimension." *New Republic,* August 1, 1983, 23–24.

Dasgupta, Gautam. "From Science to Theatre: Dramas of Speculative Thought." *Performing Arts Journal* 26–27 (1986): 237–46.

Davy, Kate. "Foreman's *Pain(T)* and *Vertical Mobility.*" *TDR* 18 (June 1974): 26–37.

——. "Richard Foreman's Theatre." *Studio International* (July / August 1976): 26–30.

——. "Kate Manheim as Foreman's Rhoda." *TDR* 20 (September 1976): 37–50.

——. "Richard Foreman's Ontological-Hysteric Theatre: The Influence of Gertrude Stein." *Twentieth Century Literature* 24 (1978): 108–26.

Falk, Florence. "Setting as Consciousness," *Performing Arts Journal* 1 (spring 1976): 51–61.

——. "Physics and the Theatre: Richard Foreman's *Particle Theory.*" *Educational Theatre Journal* 29 (October 1977): 395–404.

Gill, Brendan. "Echt Brecht." *New Yorker,* May 10, 1976, 103.

Gussow, Mel. "Celebrating the Fallen World." *New York Times,* January 17, 1994, C11.

Kauffmann, Stanley. "Ontological-Hysteric Theatre." *New Republic,* January 27, 1973, 26, 35.

Kirby, Michael. "Richard Foreman's Ontolgical-Hysteric Theatre." *TDR* 17 (June 1973): 5–32.

Leverett, James. "Old Forms Enter the New American Theatre." *New York Literary Forum* 7 (1980): 107–22.

MacDonald, Erik. "Richard Foreman and the Closure of Representation." *Essays in Theatre* 9 (1990): 19–30.

Munk, Erika. "Film Is Ego / Radio Is God: Richard Foreman and the Arts of Control." *TDR* 31 (winter 1987): 143–49.

Nahson, Edna. "With Foreman on Broadway: Five Actors' Views." *TDR* 20 (September 1976): 83–100.

Pasquier, Marie. "Richard Foreman: Comedy Inside Out." *Modern Drama* 25 (1982): 534–44.

Perec, Georges. "O images, vous suffisez à mon bonheur." *La Quinzaine Littéraire* (Paris), September 1, 1973.

Pontbriand, Chantal. "'The Eye Finds No Fixed Point on Which to Rest . . .'" *Modern Drama* 25 (1982): 154–62.

Reynaud, Bérénice. "Petite introduction à l'oeuvre de Richard Foreman." *Musique en jeu*, no. 29, November 1977.

Robinson, Marc. "A Theatre of One's Own." *Village Voice,* April 23, 1994, 92–94.

———. "Richard Foreman Loses His Head," *Theater* (Yale) 28, no. 1 (1997): 5–14.

Rockwell, John. "The Magic Theatre of Richard Foreman." *New York Times,* February 8, 1976, C11.

Savran, David. "Richard Foreman." *American Playwrights since 1845.* Edited by Philip C. Holin. New York: Greenwood Press, 1989, 102–10.

Scarpetta, Guy. "Le corps américain." *Tel Quel* (Paris), nos. 71 and 73, autumn 1977, 247–70.

———. "Richard Foreman's Scenography." *TDR* 28 (summer 1984): 23–31.

Schechner, Richard. "The Decline and Fall of the (American) Avant-Garde." *Performing Arts Journal* 14 (1981):48–63 and *Performing Arts Journal* 15 (1981): 9–19.

———. "Richard Foreman on Richard Foreman." *TDR* 31 (winter 1987): 125–35.

Shyer, Laurence. "A Night at the Opera." *Theater* (Yale) 12 (summer / fall 1981): 16–21.

Permissions

Grateful acknowledgment is made for permission to reprint from the following writers and publishers: to Arthur Bartow for a selection from "Both Halves of Richard Foreman" and to *American Theatre*, © 1987 by Theatre Communications Group, Inc.; to Kenneth Bernard for review of *Lava* and to *Studies in American Drama: 1945 to the Present, vol. 5,* © 1990 by the Ohio State University Press; to Ben Brantley for review of *My Head Was a Sledgehammer* and to the *New York Times,* copyright © 1994 by The New York Times Co., reprinted by permission; to Kate Davy for "Kate Manheim as Foreman's Rhoda" and to *TDR,* © 1976 by *TDR;* to Florence Falk for "Setting as Consciousness" and to *Performing Arts Journal,* © 1976 by *Performing Arts Journal;* to Michael Feingold for review of *Eddie Goes to Poetry City,* © 1991 by Michael Feingold, reprinted by permission of Michael Feingold and the *Village Voice;* to Richard Foreman for "Ages of the Avant-Garde," © 1994 by Richard Foreman, originally published in *Performing Arts Journal;* to Richard Foreman for "The Carrot and the Stick," © 1976 by Richard Foreman, originally published in *October;* to Richard Foreman for "How to Write a Play," © 1976 by Richard Foreman, originally published in *Performing Arts Journal;* to Richard Foreman for *Ontological-Hysteric Manifesto I,* © 1972 by Richard Foreman; to Richard Foreman and to the Overlook Press for "Samuel II," © 1994 by Richard Foreman, originally published by the Overlook Press, 2568 Rte. 212, Woodstock, NY 12498, as part of *No-Body: A Novel in Parts;* to Richard Foreman for *My Head was a Sledgehammer* and to the Overlook Press, copyright © 1995 by Richard Foreman, originally published by The Overlook Press; to Robert Gross for review of *I've Got the Shakes* and to *Theatre Journal,* © 1995 by the Johns Hopkins University Press; to Mel Gussow for "Celebrating the Fallen World" and to the *New York Times,* copyright © 1994 by The New York Times Co., reprinted by permission; to Richard Kostelanetz for excerpts from "Writing and Performance" and to *New York Arts Journal,* © 1982 by Richard Kostelanetz; to Elizabeth LeCompte for excerpts from "Off-Broadway's Most Inventive Directors Talk about Their Art," © 1997 by Elizabeth LeCompte, reprinted by permission of Elizabeth LeCompte and the *Village Voice;* to Bonnie Marranca for review of *Pandering to the Masses,* from *Theatre of Images,* New York, 1977, copyright © 1977 Bonnie Marranca, reprinted by permission of the author; to Erika Munk for review of *Film Is Evil, Radio Is Good,* © 1987 by Erika Munk, reprinted by permission of the author and the *Village Voice;* to Gerald Rabkin for review of *Don Juan* and to *Performing Arts Journal,* copyright © 1982 by *Performing Arts Journal;* to Frank Rich for review of *Penguin Touquet* and to the *New York Times,* copyright © 1981 by The New York Times Co., reprinted by permission; to Marc Robinson for "A Theatre of One's Own," © 1994 by Marc Robinson, reprinted by permission of the author and the *Village Voice;* to Gordon Rogoff for review of *The Cure,* © 1986 by Gordon Rogoff, reprinted by permission of the author and the *Village Voice;* to Arthur Sainer for review of *Total Recall,* © 1971 by Arthur Sainer, reprinted by permission of the author and the *Village Voice;* to David Savran for excerpts from "Both Halves of Richard Foreman" and to *American Theatre,* © 1987 by Theatre Communications Group, Inc.; to Guy Scarpetta for "Richard Foreman's Scenography" and to *TDR,* © 1984 by *TDR;* to Richard Schechner for review of *Rhoda in Potatoland,* © 1976 by Richard Schechner, reprinted by permission of the author and the *Village Voice;* and to Michael T. Smith for review of *Evidence* and to the *Village Voice,* © 1972 by Michael T. Smith, reprinted by permission of the author and the *Village Voice.* All material by Richard Foreman © Richard Foreman.

The following have graciously given permission to reproduce their photographs: Paula Court, © 1985, 1985, 1989, 1991, 1992, 1994, 1997, 1998; Pamela Duffy, © 1986; and Babette Mangolte, copyright © 1972, 1973, 1973, 1974, 1975, 1975, 1976, 1977, 1977, 1978, 1983, Babette Mangolte, all rights of reproduction reserved.